HEALTHCARE TECHNOLOGIES SERIES 12

Engineering High Quality Medical Software

IET Book Series on e-Health Technologies – Call for Authors

Book Series Editor: Professor Joel P. C. Rodrigues, the National Institute of Telecommunications (Inatel), Brazil and Instituto de Telecomunicações, Portugal

While the demographic shifts in populations display significant socio-economic challenges, they trigger opportunities for innovations in e-Health, m-Health, precision and personalized medicine, robotics, sensing, the Internet of Things, cloud computing, Big Data, Software Defined Networks, and network function virtualization. Their integration is however associated with many technological, ethical, legal, social and security issues. This new Book Series aims to disseminate recent advances for e-Health Technologies to improve healthcare and people's wellbeing.

Topics considered include Intelligent e-Health systems, electronic health records, ICT-enabled personal health systems, mobile and cloud computing for e-Health, health monitoring, precision and personalized health, robotics for e-Health, security and privacy in e-Health, ambient assisted living, telemedicine, Big Data and IoT for e-Health, and more.

Proposals for coherently integrated International multi-authored edited or co-authored handbooks and research monographs will be considered for this Book Series. Each proposal will be reviewed by the Book Series Editor with additional external reviews from independent reviewers. Please email your book proposal for the IET Book Series on e-Health Technologies to: Professor Joel Rodrigues at joeljr@ieee.org or joeljr@inatel.br

Engineering High Quality Medical Software

Regulations, standards, methodologies and tools for certification

Antonio Coronato

The Institution of Engineering and Technology

Published by The Institution of Engineering and Technology, London, United Kingdom

The Institution of Engineering and Technology is registered as a Charity in England & Wales (no. 211014) and Scotland (no. SC038698).

First published 2018

The Institution of Engineering and Technology
Michael Faraday House
Six Hills Way, Stevenage
Herts, SG1 2AY, United Kingdom

www.theiet.org

British Library Cataloguing in Publication Data
A catalogue record for this product is available from the British Library

ISBN 978-1-78561-248-0 (hardback)
ISBN 978-1-78561-249-7 (PDF)

Typeset in India by MPS Limited

Contents

List of Acronyms

Acronym	Test
AECL	Atomic Energy of Canada Limited
AI	Artificial Intelligence
AIMDD	Active Implantable Medical Devices
ANVISA	Agência Nacional de Vigilância Sanitária
ATDD	Acceptance Test Driven Development
BDD	Behavior Driven Development
BRH	Brazil Registration Holder
CAGR	Compound Annual Growth Rate
CAPA	Corrective Action, Preventive Action
CCB	Configuration Control Board
CE	communaute europeenne
CFDA	China Food and Drug Administration
CFR	Code of Federal Regulations
CFS	Certificate of Free Sale
CI	Configuration Item
CNS	Central Nervous System
COTS	Commercial Off-the-Shelf
CPA	Critical Path Analysis
CT	Computed Tomography
CTL	Branching Time Logic
DHF	Design History File
DPRA	Dynamic Probabilistic Risk Assessment
DTMC	Discrete-Time Markov Chain
EEA	European Economic Area
EHR	Electronic Health Record
ET	Event Tree
EU	European Union
FD&C	Federal Food Drug & Cosmetic Act
FDA	Food and Drug Administration
FMEA	Failure Mode Effects Analysis
FMECA	Failure Mode, Effects and Criticality Analysis
FT	Fault Tree
FTA	Fault Tree Analysis
FURLS	FDA Unified Registration and Listing System

GANTT	Generalized Activity Normalization Time Table
HACCP	Hazard Analysis and Critical Control Points
HAZOP	Hazard Operability Analysis
IEC	International Electrotechnical Commission
IMDRC	Import Medical Device Registration Certificate
IMDRF	International Medical Device Regulators Forum
INMETRO	Institute of Metrology, Standardization and Industrial Quality
ISO	International Organization for Standardization
IVDMD	In Vitro Diagnostic Medical Devices
JMDN	Japanese Medical Devices Nomenclature
KPI	Key Performance Indicator
LTL	Linear Temporal Logic
MAH	Marketing Authorization Holder
MD	Medical Device
MDD	Medical Device Directive
MDEL	Medical Device Establishment License
MDL	Canadian Medical Device License
MDP	Markov Decision Process
MDR	Medical Device Regulation
MHLW	Minister of Health, Labour and Welfare
MRI	Magnetic Resonance Imaging
OR	Operation Room
PAL	Pharmaceutical Affairs Law
PDA	Personal Digital Assistant
PDCA	Plan, do, check, and act
PEMS	Programmable Electrical Medical System
PERT	Program Evaluation and Review Technique
PET	Positron Emission Tomography
PHA	Preliminary Hazard Analysis
PM	Project Manager
PMA	Pre-Market Approval
PMD	Pharmaceutical and Medical Device
PMDA	Pharmaceuticals and Medical Devices Agency
PMI	Project Management Institute
PRA	Probabilistic Risk Assessment
PRM	Probabilistic Risk Model
QA	Quality Assurance
QC	Quality Control
QMS	Quality Management System
RA	Risk Assessment
RCA	Risk Control Action
RCB	Registered Certification Body
RCM	Risk Control Measure
RCO	Risk Control Operation

RCP	Risk Control Point
RFID	Radio-frequency identification
SaMD	Software as Medical Device
SDLC	Software Development Life Cycle
SFDA	State Food and Drug Administration
SOUP	Software of Unknown Provenance
SQL	Structured Query Language
SRS	Software Requirements Specification
STED	Summary Technical Document
SWOT	Strengths, Weaknesses, Opportunities and Threats
TDD	Test Driven Development
TGA	Therapeutic Goods Administration
TIR	Technical Information Report
TPD	Therapeutic Products Directorate
TQM	Total Quality Management
TR	Technical Report
UML	Unified Modeling Language
V&V	Verification&Validation
VV&T	Verification, Validation, and Testing
WBS	Work Breakdown Structure

Preface

The idea of writing a book on the quality and certification of Software as a Medical Device is, above all, the response to a professional need that I faced a few years ago.

When I started dealing with the quality of software for medical applications, my cultural background included software engineering skills and techniques and methodologies for the quality of general purpose software. On entering the healthcare domain, I promptly encountered the difficulty of finding reliable information that would allow me to quickly obtain an insight into the industry's regulations, processes, requirements and practices. In fact, I did not find a 'single starting point' from which to begin figuring out the problem.

Moreover, I encountered two further difficulties.

The first is that much of the documentation, manuals, regulations and information available on the network refers to traditional medical devices; i.e., equipment whose architecture is mainly based on hardware components where the role of the software (if any) is absolutely marginal. Instead, I was targeting complex, distributed software systems that needed to integrate with other heterogeneous software systems. But software, by its nature, has peculiarities that make it clearly different from any other manufacturing product, including even traditional Medical Devices. Software does not wear out but deteriorates. A software product is the result of one-time execution of an ad-hoc development cycle. So, the consolidated practices for the development of traditional Medical Devices and the manuals that describe them have a limited utility and may even be misleading in some cases.

The other difficulty concerns the diversity of markets. The regulations and standards that apply to the European market are certainly not suitable to other markets. Indeed, they may be incompatible with the requirements of these latter.

Consequently, this book aims to provide a 'starting point' and offer a rather comprehensive overview of issues related to the production and marketing of Software as a Medical Device. Some of the issues discussed, such as Project Management, Quality Management and Risk Management, would require specific insights, but the idea is to provide the right pointers from which, once the general issue is clear, the reader can extend his/her degree of knowledge.

Part I

Introduction

Chapter 1
Introduction

1.1 The evolution of medical purpose software

In the health-care domain, over the years, software has become a pervasive technology, no longer confined to controlling traditional **medical devices** (e.g., radiology equipment).

Medical purpose software has been assuming increasing responsibilities by redefining classic health-care services or offering new facilities. It is now found on desktop and mobile devices to give health-care operators access to clinical information stored in large and distributed applications of electronic health records or to clinical decision support systems, wherever and whenever they need. Modern hospitals are, nowadays, supported by a variety of complex software systems able to ease and improve almost all clinical practices. For example, advanced operation rooms (OR), also known as intelligent ORs, are "context-aware" smart spaces able to automatically detect which step of the procedure the OR team is at and to identify procedural delays and deviations from standard protocols. Some institutions are experimenting with robot-assisted tele-surgery with surgeons remotely located in immersive virtual reality environments. Chronic patients may now experience a better quality of life, far from the hospital, thanks to tele-monitoring applications that continuously analyze their vital signs and autonomously predict the approach of critical conditions. Applications of ambient-assisted living have turned the patient's home into an intelligent environment able to provide facilities to support daily activities and to collect information useful to both better assess the evolution of the disease and to establish a proper treatment.

These are just a few examples of new classes of medical purpose software (software used to make clinical decisions) all of which—with a different "level of concern"—must be safe, dependable, reliable, and secure. A certain level of quality, of course, may be a desired characteristic that medical purpose software producers would like to confer on their products in order to better compete on the market. However, this is not enough. Indeed, some of the previously described systems must be treated as medical devices themselves (referred as **software as medical device (SaMD)** or **medical device software**) and, consequently, they must be designed, built, and maintained according to stringent regulations and standards. The scenario is quite complex due to the following main elements:

1. The emergence and diffusion of new technologies have opened up tremendous opportunities for market players and stakeholders (e.g., mobile and wearable

technologies) and have extended the category of medical purpose software that is no longer confined to small control programs of medical equipment (e.g., Assembler/C routines embedded in a medical device), but includes a variety of software systems much wider and heterogeneous than that of traditional medical devices;
2. The existence of stringent regulations for medical devices, which may differ country-by-country and must be taken into account when designing SaMD;
3. The existence of several standards and guidelines that are suggested by the regulations in order to satisfy some of the requirements posed by the regulations themselves;
4. High-quality is a desired characteristic for medical purpose software, whereas it is a de facto requirement for SaMD, which must be designed and maintained in compliance with medical-device regulations.

It must be considered also that regulations were generally conceived some years ago and focused on the classic vision of the medical device, but recently the International Medical Device Regulators Forum (IMDRF) has taken responsibility for the issue of establishing a common framework for regulators to incorporate converged controls into their regulatory approaches for SaMD.

First of all, the IMDRF has provided a definition for SaMD as "software intended to be used for one or more medical purposes that perform these purposes without being part of a hardware medical device" [1], to differentiate it from **software in a medical device** (sometimes referred as "embedded" or "part of"), which includes the software control functions of traditional medical equipment.

The aim of this book is to explore this complex scenario and to illustrate how to exploit techniques, methodologies, development processes, and existing standards to realize high-quality medical purpose software, according to the regulations. The certification of SaMD is the goal of the engineering tasks illustrated in the book.

1.2 Product quality and software quality

According to the American Association for Quality, **quality** is "a subjective term for which each person or sector has its own definition. In technical usage, quality can have two meanings: (1) the characteristics of a product or service that bear on its ability to satisfy stated or implied needs; (2) a product or service free of deficiencies."

Quality is strictly related to the ability of the product to meet given requirements and customer expectations. From the point of view of the client, the quality of prime materials, color, or comfort does not matter. What matters is that the client receives what he expects when buying that product. Thus, the closer the product is to the expectations, the better the quality of the product is perceived. On the other hand, the manufacturer is primarily concerned with the design, engineering, and production processes. From his point of view, quality is measured by the degree of conformance to predetermined specifications and standards. Deviations from such specifications and standards are identified as product defects, which lead to poor quality and low reliability, and efforts for their remotion are required in order to improve the product quality.

It may be surprising to discover that one of the first documents related to product quality was found in a paper discovered in an Egyptian archeological site (almost 1450 B.C.). The paper reported measurements for bricks accepted for building monuments. An activity of dimension checking was in place to discard noncompliant bricks.

During the Second World War, the necessity of controlling the effectiveness of weapons before their release emerged. It was with respect to this application domain that **quality-control** (QC) tasks were engineered and disseminated all along the manufacturing processes. Such activities had the unique objective of checking the conformity of the product against specifications and discarding those below a certain level of quality called the "acceptable quality level." A few years later, after the end of the Second World War, statistical methods were developed for both quality and process control.

A major shift, however, took place when the focus was moved from QC to **quality assurance**. The goal now was to find strategies, techniques, and tools that allow the increment of manufacturing processes in order to reduce the amount of imperfect products that should be discarded by QC activities. **Total quality management** (TQM) was the new approach devised to improve quality through planning and executing manufacturing and control processes throughout an organization. TQM describes a management approach to long-term success through customer satisfaction. It is substantially a management system that allows to put the focus on the customer and involves all employees in continual improvement. It relies on strategy, data, and effective communications to integrate the quality culture into core activities of the organization. The following are the eight principles of TQM:

1. **Customer-focused.** As long as a business organization depends on its customers, the effort to improve the product quality starts with the understanding of their needs and expectations. Customers ultimately determine the level of quality—whatever the organization does to foster quality improvement—and customers determine whether the efforts are effective.
2. **Total employee involvement.** All employees are involved and participate in working toward common goals. A good quality culture needs to be fostered before achieving high-quality levels.
3. **Process-centered.** A central aspect of the TQM approach is the focus on processes. A process is a predefined sequence of phases interconnected and communicating with each another. Performances are continuously monitored in order to detect deviations or inefficiencies.
4. **Integrated system.** Although an organization may consist of many vertical departments, TQM upholds such vertical structures in an integrated fashion by means of horizontal processes.
5. **Strategic and systematic approach.** TQM requires a strategic and systematic approach that integrates quality as a core component of the organization's vision, mission, and goals.

6. **Continual improvement**. A key principle of TQM is continual process improvement. One way to aim at continual improvement is set and execute specific actions like those defined by the PDCA (plan, do, check, and act) cycle: plan, do, check, and act. It is a four-stage process executed iteratively. In the plan phase, the goal to achieve or the problem to solve is defined in the do phase, a solution is realized and applied; in the check phase, the effectiveness of the solution is evaluated; and in the act phase, countermeasures are eventually taken.
7. **Fact-based decision-making**. TQM requires that the organization continually collects and analyzes data about the effectiveness of its processes in order to improve decision-making accuracy and allow predictions based on past history.
8. **Communications**. Effective communications play a remarkable role in maintaining in motivating employees at all levels. Communications involve strategies, method, and timeliness.

Many of these concepts are present in modern **quality management systems** (QMSs), the successor to TQM, which is discussed later in this book.

TQM and QMS principles are applicable to all kinds of organization and product, but it is important to note that a software system has features that make it special with respect to other products. It is not the result of a classic manufacturing process (e.g., an assembly line in a factory). A software system is immaterial, and the quality does not deteriorate with use in time, but with maintenance operations. As a consequence, the question "What is software quality?" is likely to be connected with responses different from other kinds of product and is, generally, difficult to answer. An alternative question is: "What are the characteristics of quality software?", with respect to the different quality views (customer and manufacturer) and expectations. From this perspective, some models and frameworks have been proposed to define quality-related attributes and measurements. The most common is the ISO-9126 framework [2].

ISO-9126 defines a two-level hierarchical structure of software quality attributes. The top level is composed of six macroattributes (external attributes), each one connected to subattributes (internal attributes). Every subattribute is connected exclusively to one macro-attribute. The list of attributes is as follows:

Functionality (connected to *suitability*, *accuracy*, *interoperability,* and *security*): The capability of the software product to provide functions which meet stated and implied needs when the software is used under specified conditions.
Reliability (connected to *maturity*, *fault tolerance,* and *recoverability*): The capability of the software product to maintain a specified level of performance when used under specified conditions.
Usability (connected to *understandability*, *learnability,* and *operability*): The capability of the software product to be understood, learned, used, and attractive to the user, when used under specified conditions.
Efficiency (connected to *time behavior* and *resource behavior*): The capability of the software product to provide appropriate performance, relative to the amount of resources used, under-stated conditions.

Maintainability (connected to *analyzability*, *changeability*, *stability,* and *testability*): The capability of the software product to be modified. Modifications may include corrections, improvements, or adaptation of the software to changes in the environment, and in requirements and functional specifications.
Portability (connected to *adaptability*, *installability*, *conformance,* and *replaceability*): The capability of the software product to be transferred from one environment to another.

Software product quality can then be evaluated by measuring the internal attributes (the metrics are defined in ISO IEC 9126 Part III), or by measuring external attributes (the metrics are defined in ISO IEC TR 9126 Part III), or by measuring the quality in use attributes (the metrics are defined in ISO IEC TR 9126 Part III) (Figure 1.1)

1.3 On the need for quality in medical purpose software

The need for engineering safety and quality for medical purpose software emerged dramatically after the Therac-25 disaster. Therac-25 was a radiation therapy machine produced by Atomic Energy of Canada Limited (AECL) as an evolution of a previous model. Such a medical appliance was involved in at least six documented accidents between 1985 and 1987 that caused death or severe injury to patients because of massive radiation overdoses received during the examination.

A commission established that the accidents were caused by different software defects and usability issues. The investigation reported that the primary reason should be attributed to the bad software design and development practices, which caused the introduction of several defects in the software code.

As reported by Leveson and Turner [3], overconfidence in the ability of software to ensure the safety of the Therac-25 was an important factor which led to the accidents.

The risk assessment was unrealistic and substantially focused on hardware failure modes. Moreover, the follow-up on the earlier accidents was superficial and unacceptable. In response to one reported accident, the manufacturer tried to reproduce the condition which had occurred, but the engineers were not capable of doing this. As a result, they concluded that the accident had been caused by a hardware fault and implemented a solution that, however, did not stop accidents. The Therac-25 incidents demonstrate that several misconceptions led to the accidents: overconfidence in the software's abilities, poor software design and verification, and unreasonably low-risk assessments.

Manufacturers have to understand that rigorous testing and failure analyses are essential for such a kind of software system.

Many lessons can be learned from this series of accidents [4,5]:

Overconfidence in software—A common mistake is to put too much confidence in the software, especially among nonsoftware professionals who seem to believe software will not or cannot fail. The first safety analysis on the Therac-25 did not include software although major responsibilities for safety rested on it. When problems began to be reported, technicians from AECL assumed that hardware faults were the cause, and the investigation was mainly conducted at the hardware layer.

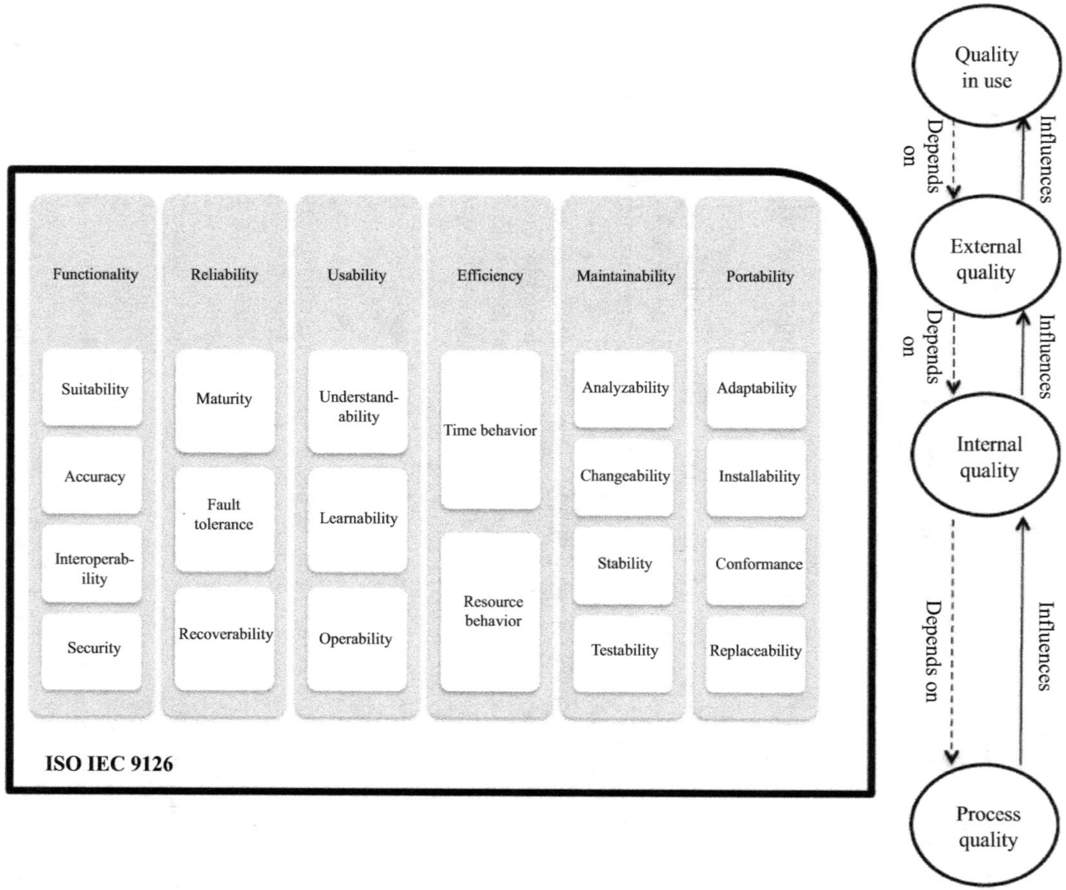

Figure 1.1 ISO IEC 9126

Confusing reliability with safety—This software was highly reliable. It worked tens of thousands of times before overdosing anyone, and occurrences of erroneous behavior were few. AECL assumed that their software was safe because it was reliable, and this led to complacency.

Lack of defensive design—The software did not contain self-checks or other error-detection and error-handling features that would have detected the inconsistencies and coding errors. Audit trails were limited because of a lack of memory. However, today larger memories are available, and audit trails and other design techniques must be given a high priority in making trade-off decisions. Patient reactions were the only real indications of the seriousness of the problems with the Therac-25. There were no independent checks that the machine and its software were operating correctly. Such verification cannot be assigned to operators without providing them with some means of detecting errors. The Therac-25 software lied to the operators, and the machine itself was not capable of detecting that a massive overdose had occurred. Engineers have to design for the worst case.

Failure to eliminate root causes—One of the lessons to be learned from the Therac-25 experience is that focusing on particular software design errors is not the way to make a system safe. Virtually, all complex software can be made to behave in an unexpected fashion under some conditions: There will always be another software bug. Just as engineers would not rely on a design with a hardware single point of failure that could lead to catastrophe, they should not do so if that single point of failure is software.

One of the serious mistakes that led to the multiple Therac-25 accidents was the tendency to believe that the cause of an accident had been determined without adequate evidence to come to this conclusion and without looking at all possible contributing factors. Without a thorough investigation, it is not possible to determine whether a sensor provided the wrong information, the software provided an incorrect command, or the actuator had a transient failure and did the wrong thing on its own.

In general, it is a mistake to patch just one causal factor such as the software and assume that future accidents will be eliminated. Accidents are unlikely to occur in exactly the same way again. If we patch only the symptoms and ignore the deeper underlying causes, or if we fix only the specific cause of one accident, we are unlikely to have much effect on future accidents. The series of accidents involving the Therac-25 is a good example of exactly this problem: fixing each individual software flaw as it was found did not solve the safety problems of the device.

Complacency—Often, it takes an accident to alert people to the dangers involved in technology. A medical physicist wrote about the Therac-25 accidents: "In the past decade or two, the medical accelerator industry has become perhaps a little complacent about safety. We have assumed that the manufacturers have all kinds of safety design experience since they have been in the business a long time. We know that there are many safety codes guides and regulations to guide them, and we have been reassured by the hitherto excellent record of these machines. Except for a few incidents in the 1960s, the use of medical accelerators has been remarkably free of serious radiation accidents until now; perhaps, though we have been spoiled by this success."

Unrealistic risk assessments—The first hazard analyses initially ignored software and then they treated it superficially by assuming that all software errors were equally likely. The probabilistic risk assessments generated undue confidence in the machine and in the results of the risk assessment themselves. When the first Yakima accident was reported to AECL, the company did not investigate. Their evidence for their belief that the radiation burn could not have been caused by their machine included a probabilistic risk assessment showing that safety had increased by orders of magnitude as a result of a fix. The belief that safety had been increased by such a large amount seems hard to justify. The problem with all such analyses is that they typically make many independence assumptions and exclude aspects of the problem that are difficult to quantify but which may have a larger impact on safety than the quantifiable factors that are included.

Inadequate investigation or follow up on accident reports—Every company building safety critical systems should have audit trails and incident analysis procedures that are applied whenever any hint of a problem is found that might lead to an accident. The first phone call reporting the news of the first accident should have led to an extensive investigation of the events at the scene of the event.

Inadequate software engineering practices—Some basic software engineering principles that apparently were violated in the case of the Therac-25, which include the following: (1) software specifications and documentation should not be an afterthought; (2) rigorous software quality assurance practices and standards should be established; (3) designs should be kept simple and dangerous coding practices avoided; (4) ways to detect errors and get information about them, such as software audit trails, should be designed into the software from the beginning; (5) the software should be subjected to extensive testing and formal analysis at the module and software level; system testing alone is not adequate. Regression testing should be performed on all software changes; and (6) computer displays and the presentation of information to the operators, such as error messages, along with user manuals and other documentation need to be carefully designed

Software reuse—Important lessons about software reuse can be found in these accidents. A naive assumption is often made that reusing software or using commercial off-the-shelf software will increase safety because the software will have been exercised extensively. Reusing software modules does not guarantee safety in the new system to which they are transferred and sometimes leads to awkward and dangerous designs. Safety is a quality of the system in which the software is used, it is not a quality of the software itself. Rewriting the entire software in order to get a clean and simple design may be safer in many cases.

Safe versus friendly user interfaces—Making the machine as easy as possible to use may conflict with safety goals. Certainly, the user interface design left much to be desired but eliminating multiple data entry and assuming that operators would check the values carefully before pressing the return key was unrealistic.

The Food and Drug Administration (FDA) got involved in the Therac-25 incidents, and the response was impressive especially considering how little experience they had with similar problems in computer controlled medical devices. Since the Therac-25 events, the FDA has moved to improve the reporting system and to augment their procedures and guidelines to include software. In response to incidents like those associated with Therac-25, the IEC 62304 standard was created. It introduces development life cycle standards for medical device software and specific guidance on using software of unknown pedigree.

1.4 Regulatory environments

SaMD, as well as embedded software (software in a medical device), falls under the regulatory environment.

Many countries and economic communities have developed regulatory frameworks that pose stringent requirements in order to place on their market a medical device. A regulatory framework establishes different classes for medical devices depending on their level of concern. The number of classes ranges, framework-by-framework, from three up to five. The classification system is either risk based or rule based. The higher the medical device's risk class, the more requirements the manufacturer must meet to get license approval.

The requirements concern four major topics:

1. The QMS
2. The risk management
3. The system verification and validation

The manufacturer has to provide evidence that all requirements are met. Competent authorities then check such evidence during the approval process and eventually clear the device, which successively can be placed on the market.

The regulatory environment in the European Union relies on three directives respectively for active implantable devices, in vitro diagnostic devices, and all other medical devices. Directive 93/42/EEC and its successive modification, Directive 2007/47/EC, regard medical devices and medical purpose software. The directive defines five classes of medical devices: "Class I (nonsterile, nonmeasuring)," "Class I (sterile or measuring)," "Class IIa," "Class IIb," and "Class III." The manufacturer prepares a "declaration of conformity" and affixes the CE marking after a "notified body" has audited both its QMS and the device "technical file" ("design dossier" for Class III devices) successfully. In the case of Class I (nonsterile, nonmeasuring) devices, no audit is required. Instead, the manufacturer has just to register the device with a notified body.

Medical devices sold within the United States are subject to the Title 21 Code of Federal Regulations Part 800-1200 (21 CFR Parts 800-1299). In particular, 21 CFR Part 820 concerns setting company policies, operating procedures, and objectives for a quality system. The regulation establishes three classes of devices: "Class I" (minor concern), "Class II" (moderate concern), and "Class III" (major concern). The classification system is risk based. Once again, different approval pathways take place depending on the device class. Devices in Class III need premarket approval and undergo specific inspections performed by the FDA. For devices falling in Class II, a 510-K premarket submission is required, but FDA inspections are not mandatory.

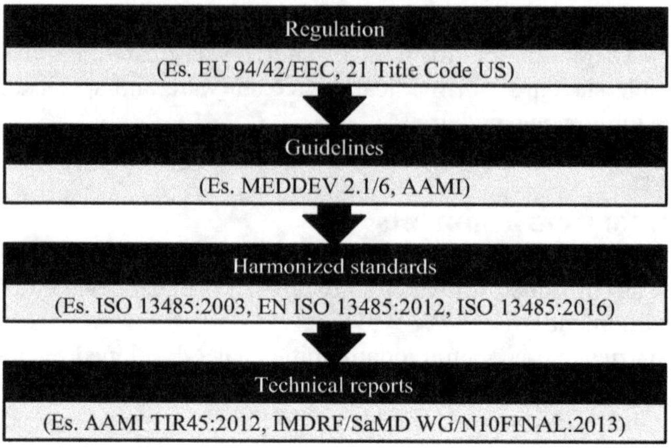

Figure 1.2 Normative documents stack

Part I of this book presents many more details related to regulations in the USA, Europe and other countries and economic communities. It is important to note, however, that regulations stand on top of a stack of normative documents as shown in Figure 1.2. All such layers are extremely relevant for the production of high-quality medical purpose software.

Guidelines are official government documents that provide further information or clarify the application of the law. As an example, the MEDDEV 2.1/6 ("Guidelines on the Qualification and Classification of Stand-Alone Software used in Healthcare within the Regulatory Framework of Medical Devices") concerns the problem of qualifying (i.e., establishing if the system under examination is a medical device) and classifying (i.e., executing the rule-based decision algorithm to determine the class of concern) a medical purpose software in the European regulatory environment. The guideline introduces several examples to show the application of the rules and the qualification and classification result. Guidelines are not prescriptive documents; their importance is related to the correct comprehension and application of the law.

The layer of standards provides requirements, specifications, and guidelines that can be used consistently to ensure that the medical purpose software under construction will meet a part of the requirements posed by the regulations. In other words, some standards have been recognized as they are, or harmonized, by regulations to match requirements. As an example, ISO 13485:2003 (international standard for "medical devices—QMSs—requirements for regulatory purposes") (1) is not recognized by the FDA that has its own set of requirements for the QMS; (2) has been harmonized by European authorities in the EU ISO 13485:2012; and (3) has been recognized by Canadian authorities as it is. Consequently, a manufacturer displaying an ISO 13485:2003 compliant QMS will have easier access to the market in Canada than in Europe, the more difficulties in the USA.

The technical reports (TRs) layer collects publications that address a particular aspect or issue. A TR may be valuable for industries and practitioners although it is not a standard, which, contrary to a TR, is subject to a formal process of review and approval. A TR is not subject to the same formal approval process as a standard, but it is approved for distribution by a technical committee. AAMI TIR45: 2012 ("Technical Information Report Guidance on the use of AGILE practices in the development of medical device software") is a relevant example of a TR that reports on the use of agile methodologies for the development of medical purpose software.

With the exception of the regulatory layer, all other layers concern nonmandatory prescriptions for the manufacturer, although these are highly recommended.

1.5 Verification and validation

On January 2002, the FDA released a document titled General Principles of Software Validation that reports general validation principles applicable to the case of medical device software [6]. These principles apply to (1) software used as a component, part, or accessory of a medical device, (2) software that is itself a medical device, and (3) software utilized in the production of a device. In general, quality expectations for software systems are 2-fold: (1) they must perform their operations correctly or satisfactorily (i.e., they must do the "things right") and (2) the software systems must do what they are supposed to do by meeting the intended use (i.e., they must do the "right things"). The former requires that the software system operates properly. It must implement the requirements correctly. The latter requires that the software system performs only the intended functions. It must fulfill the right requirements.

While **verification** is the term adopted to indicate tasks performed to demonstrate that the system does the things right, **validation** activities are those intended to confirm that the system does the right things, whereas **testing**, generally speaking, concerns techniques, activities, processes, policies, and strategies focused on measuring the quality of a software system. Some people refer to verification, validation, and testing as if it was a single concept.

As reported by the General Principles of Software Validation document: "Software verification provides objective evidence that the design outputs of a particular phase of the software development life cycle meet all of the specified requirements for that phase. Software verification looks for consistency, completeness, and correctness of the software and its supporting documentation, as it is being developed, and provides support for a subsequent conclusion that the software is validated. Software testing is one of the many verification activities intended to confirm that software development output meets its input requirements. Other verification activities include various static and dynamic analyses, code and document inspections, walkthroughs, and other techniques. Software validation is intended "to be a confirmation by examination and provision of objective evidence that software specifications conform to user needs and intended uses, and that the particular requirements implemented through software can be consistently fulfilled."

Software verification activities may take place during the development cycle, as well as at the end of the process, to ensure that the requirements have been implemented correctly and are traceable.

Software verification and validation are based on comprehensive software testing, inspections, analyses, formal verification, and other verification tasks performed during the software development life cycle. Software testing should be performed both in a simulated use environment and at the user site under real operating conditions.

Verification and validation activities are difficult to manage because it is impossible to achieve the exhaustive testing of a nontrivial system. The question is: how much evidence (testing) is sufficient? Consequently, software validation is a matter of building a "level of confidence" that the system meets requirements and user expectations, and it is free from unacceptable risks (i.e., it is safe).

Measures such as defects found, estimates of defects remaining, testing coverage, etc. are needed to develop an acceptable level of confidence before delivering the product. The level of confidence and the extent of software testing required will vary depending upon the safety risks connected to the use of the system.

1.6 Structure of the book

Chapter 1 introduces the topics covered by the book and the introductory concepts. It clarifies what is medical device software, what is not, and emphasizes the differences between the new categories of medical device software and the classic medical devices. It also introduces regulations, standards, and practices. The need for medical software compliance is explained.

Part I of the book is dedicated to regulations. In particular, Chapter 2 illustrates the regulations active in Europe (e.g., EU MDD 93/42/EEC and its modification in EU MDD 2007/47/EC). This shows the structure of the law, the medical device qualification, the classification and certification process, and all the standards recognized as adequate to respond to the regulation requirements.

Chapter 3 presents the regulations active in the USA (e.g., the FDA Title 21 of the US CFR). This shows the structure of the law, the medical device classification and certification process, all standards recognized as adequate to respond to regulation requirements, and some guidance.

Chapter 4 illustrates briefly the regulations for other main markets and countries (China, Japan, Australia, Brazil, etc.).

Part II of the book presents an overview of several standards.

Chapter 5 introduces the standard ISO 13485. It presents the structure of the standard, the quality system requirements of various global regulations, and the use of ISO 13485 in each jurisdiction.

Chapter 6 describes the standard ISO 14971. It introduces the concepts of risk, residual risk, harm, hazard, and hazardous situations. It presents the risk management process described by the standard. Several risk management methods and tools are also introduced.

Chapter 7 deals with the standard IEC 62304. It reports the main requirements for a software development life cycle. It presents a development processes and some characteristics of a maintenance processes that the standard defines with respect to the different classes of medical software. It finally reports the use of such a standard in different jurisdictions.

Chapter 8 presents other standards that may affect medical software development such as those for software verification.

Part III is dedicated to verification and validation techniques.

Chapter 9 illustrates static testing techniques (reviews, walkthroughs, and inspections) useful for any artifact of the development process. It describes the roles and techniques to perform static testing. In addition to this, the chapter illustrates the static testing techniques for source code (i.e., static analysis).

Chapter 10 deals with dynamic testing techniques. It describes both black-box (equivalence partitioning, boundary value analysis, and state transition) and white-box (statement coverage, decision coverage, and path coverage) techniques.

Chapter 11 describes some formal methods and tools useful for the specification and stringent verification of particularly critical system functions. The chapter concludes with guidelines for the application of the different verification and validation techniques with respect to different classes of requirements and systems.

Part IV discusses methods, methodologies, and engineering tasks for development, configuration, and maintenance

Chapter 12 illustrates classic models of software development life cycle. Different models (i.e., waterfall, v-model, incremental, iterative, and evolutive) are analyzed from the perspective of the manufacturer. The advantages and disadvantages of each model with respect to the case studies are discussed.

Chapter 13 introduces agile methods. It presents the agile development manifesto. It then illustrates an agile framework, namely scrum, and the test-driven development process. Pros and cons of such processes are discussed.

Chapter 14 illustrates techniques and tools for planning activities and managing responsibilities. It describes the definition of roles and the traceability.

Chapter 15 presents an integrated approach to risk management. Both static and dynamic risk analysis techniques are discussed.

Chapter 16 illustrates techniques and tools for the specification of system requirements. Semiformal methods are adopted (e.g., Unified Modeling Language models).

Chapter 17 presents design controls, which are the regulatory requirements for design activities. Several templates for implementing such controls are presented.

Chapter 18 deals with test management and defect management. It introduces some test principles and subsequently a test process. A defect management process is also shown.

Chapter 19 presents a change management process, a configuration process, and an incident management process.

Part V concludes the book with Chapter 20 that reports considerations on the "gray area," that related to new classes of software systems not yet regulated or not clearly identifiable as a medical device. It also illustrates opportunities and critical issues in this area.

Part II

Regulations

Chapter 2
EU MDD 93/42/EEC

2.1 Background

Council Resolution 85/C 136/01 of the May 7, 1985 defined the so-called "new approach" to technical harmonization and standards with the aim of removing barriers to trade and facilitating the free movement of goods within the European Union (EU). In the health-care domain, the effect of such a resolution had been the definition of a legal framework for medical devices that consists of three directives: (1) Council Directive 90/385/EEC on Active Implantable Medical Devices (1990) [7]; (2) Council Directive 93/42/EEC on Medical Devices (1993) [8]; and (3) Council Directive 98/79/EC on In Vitro Diagnostic Medical Devices (IVDMD) (1998) [9]. Therefore, such a legal framework had been originally conceived in order to "remove barriers to trade and facilitate the free movement of medical devices." The effort of harmonization, however, had resulted in the definition of a set of common requirements for safety, which had previously been defined separately by the Member States.

At present, the legal framework applies across one of the largest medical device markets in the world, the European Economic Area, which includes the 27 member states of the EU, as well as Norway, Iceland, and Liechtenstein.

Despite the requirement to transpose these directives into law in each Member State, the binding laws set by national authorities present some differences in levels and approaches between the Member States. As a result, a variety of interpretations can be observed across the EU. On the other hand, these three directives have been supplemented over time by several modifying and implementing directives, including the last technical revision introduced by Directive 2007/47/EC [10].

These documents undergo continuous development in order to reflect the state of the art in the quickly developing area of Medical Devices. In 2008, the EU started a recasting process that aims at reorganizing the directives in a medical device regulation.

In 2012, the commission decided to adopt a package of measures on innovation in health that includes a communication and two regulation proposals to revise the existing legislation on medical devices. The aim is to ensure (1) a consistently high level of health and safety protection for people using these products; (2) the free and fair trade of the products throughout the EU; and (3) that EU legislation is adapted to the significant technological and scientific progress in this sector over the last 20 years.

The commission has also issued an impact assessment on the revision of the regulatory framework for medical devices. Such a study (1) describes the problems encountered under the existing legislation; (2) justifies the need and the objectives of the Commission initiative; and (3) compares the different policy options and their impact on the medical devices sector.

For the rest of the book, we refer to the Directive 93/42/EEC, its revision (Directive 2007/47/EC), and the related guidance documents, which are currently in force.

2.2 Content of the Directive 93/42/EEC

The Directive 93/42/EEC consists of 23 articles and 12 annexes. The directive has been amended by successive directives and regulations.

Table 2.1 reports the list of articles of the directive. In the following, we will describe some of the content of the amended and consolidated version of the directive.

The first article defines the scope. It establishes that the directive shall apply to "medical devices" and their "accessories." For the purposes of the directive, accessories are treated as medical devices in their own right. Medical devices, accessories and other terms relevant to our discussion are defined in Article 1. The article also defines products to which the directive does not apply (e.g., in vitro diagnostic devices, active implantable devices covered by Directive 90/385/EEC, etc.).

Article 3 specifies that any device must meet the essential requirements set out in Annex I which are applied taking into account the intended purpose of the device concerned. The article, after the amendment of the Directive 2007/47/EC, also directly refers to hazard and safety requirements.

The classes of concern for medical devices are introduced by Article 9, which groups products into Classes I (nonmeasuring and nonsterile), I measuring, I sterile, IIa, IIb, and III. The classification is carried out in accordance with Annex IX, which reports a detailed set of eighteen classification rules.

The *intended purpose* of the device drives the application of the rules. If the device is built to be used in combination with another device, the rules act separately on both devices. Accessories are classified in their own right independently of the device they are associated with.

Software embedded in a device falls automatically into the same class as the device itself. If the device is not intended to be used in a specific part of the body, it must be evaluated and classified taking into account the most critical specified use. If several rules apply to the same device, you have to refer to the strictest one resulting in the higher classification.

The system of rules relies on the criteria shown in Table 2.2 and defines different classification routes depending on such criteria. For example, Rule 7 establishes that "surgically invasive" devices intended for "short-term use" are generally classified as Class IIa devices but, some exceptions apply (e.g., devices intended either specifically to control, diagnose, monitor or correct a defect of the heart or of the central circulatory system through direct contact with these parts of the body, are in Class III).

Table 2.1 Structure of the Directive 93/42/EEC

Article	
1	Definitions, scope
2	Placing on the market and putting into service
3	Essential requirements
4	Free movement, devices intended for special purposes
5	Reference to standards
6	Committee on standards and technical regulations
7	–
8	Safeguard clause
9	Classification
10	Information on incidents occurring following placing of devices on the market
11	Conformity assessment procedures
12	Particular procedure for systems and procedure packs and procedure for sterilization
12a	Reprocessing of medical devices
13	Decisions with regard to classification and derogation clause
14	Registration of persons responsible for placing devices on the market
14a	European data bank
14b	Particular health monitoring measures
15	Clinical investigation
16	Notified bodies
17	CE marking
18	Wrongly affixed CE marking
19	Decision in respect of refusal or restriction
20	Confidentiality
20a	Cooperation
21	Repeal and amendment of directives
22	Implementation, transitional provisions
23	–
Annex	
I	Essential requirements
II	EC declaration of conformity
III	EC type-examination
IV	EC verification
V	EC declaration of conformity
VI	EC declaration of conformity
VII	EC declaration of conformity
VIII	Statement concerning devices for special purposes
IX	Classification criteria
X	Clinical evaluation
XI	Criteria to be met for the designation of notified bodies
XII	CE marking of conformity

Article 11 regulates the conformity assessment, that is the method by which the manufacturer demonstrates that the device complies with the requirements of the directive. Of course, the classification of the medical device has a major impact on the conformity assessment route that the manufacturer should follow to demonstrate compliance and to affix the CE marking on the medical device. However, different

Table 2.2 Parameters for the classification of medical devices

Duration	
Transient	Normally intended for continuous use for less than 60 min
Short term	Normally intended for continuous use for not more than 30 days
Long term	Normally intended for continuous use for more than 30 days
Invasiveness	
Invasive device	A device which, in whole or in part, penetrates inside the body, either through a body orifice or through the surface of the body
Body orifice	Any natural opening in the body, as well as the external surface of the eyeball, or any permanent artificial opening, such as a stoma
Surgically invasive device	An invasive device which penetrates inside the body through the surface of the body, with the aid of or in the context of a surgical operation
Reusable surgical instrument	Instrument intended for surgical use by cutting, drilling, sawing, scratching, scraping, clamping, retracting, clipping, or similar procedures, without connection to any active medical device and which can be reused after appropriate procedures have been carried out
Implantable device	Any device which is intended: (1) to be totally introduced into the human body or (2) to replace an epithelial surface or the surface of the eye; by surgical intervention which is intended to remain in place after the procedure. Any device intended to be partially introduced into the human body through surgical intervention and intended to remain in place after the procedure for at least 30 days is also considered an implantable device
Critical anatomical locations	Related to the central circulatory system, or the central nervous system
Active medical devices	Any medical device operation of which depends on a source of electrical energy or any source of power other than that directly generated by the human body or gravity and which acts by converting this energy Stand-alone software is considered to be an active medical device
Devices with a measuring function	Information on devices with a measuring function can be found in MEDDEV 2.1/5
Procedure packs	Procedure packs normally do not require classification

routes may be followed by a manufacturer to provide evidence of its conformity. For example, in the case of a device falling within Class III, the manufacturer shall either: (1) follow the procedure relating to the EC declaration of conformity set out in Annex II (full quality assurance) or (2) follow the procedure relating to the EC type-examination set out in Annex III, coupled with: (i) the procedure relating to the EC verification set out in Annex IV or (ii) the procedure relating to the EC declaration of conformity set out in Annex V (production quality assurance). All these routes may lead the manufacturer to the affixing of the CE marking. The decision regarding the conformity assessment route is determined by the classification, expenses, and speed to market.

Only Class I devices (except for sterile or with measuring function) are self-declared by the manufacturer without the involvement of a **notified body**, which is an organization designated by a Member State to carry out the tasks pertaining to the conformity assessment (Article 16).

The CE marking is regulated by Article 17, as well as the wrong affixing of the CE marking by Article 18.

Annex I concerns the "essential requirements" for a medical device. The requirements are grouped into "general requirements" and "requirements regarding design and construction." The driving concept is that the device must be safe, that is free from unacceptable risks for the safety of patients, or the safety and health of users or other people. The solutions adopted by the manufacturer for the design and construction must conform to the following safety principles: (1) they must eliminate or reduce risks as far as possible (inherently safe design and construction); (2) where appropriate they must take adequate protection measures including alarms if necessary, in relation to risks that cannot be eliminated; and (3) they must inform users of the residual risks due to any shortcomings of the protection measures adopted.

The EC declaration of conformity is the theme of Annex II. The declaration of conformity is the procedure whereby the manufacturer ensures and declares that the products concerned meet the requirements posed by the directive. In the case of Class III products, a full quality assurance system must be audited in order to obtain the CE marking. The auditing must be supported by registered documents and records concerning all the facts, elements, and requirements related to the quality and safety of the product. The quality system shall include in particular an adequate description of: (1) the manufacturer's quality objectives; (2) the organization of the business, the organizational structures, the responsibilities of the managerial staff and their organizational authority, and the methods of monitoring the efficient operation of the quality system; (3) the procedures for monitoring and verifying the design of the products, including the corresponding documentation, a general description of the product and its intended use, the design specifications, the techniques used to control and verify the design and the processes and systematic measures which will be used when the products are being designed; (4) the inspection and quality assurance techniques at the manufacturing stage; and (5) the appropriate tests and trials which will be carried out before, during and after manufacture.

In addition to the obligations imposed regarding the quality system, the manufacturer must lodge with the notified body an application for the examination of the

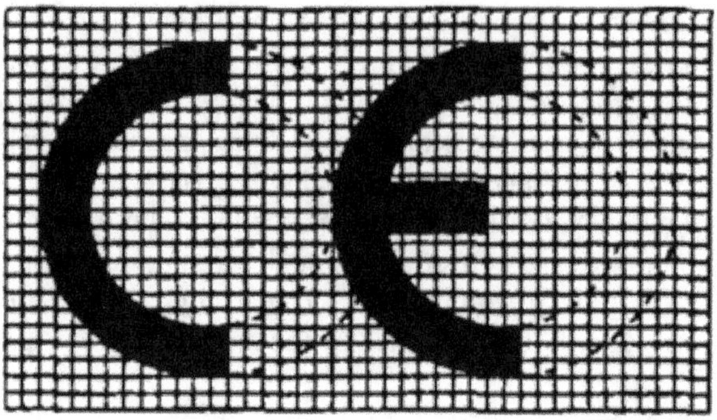

Figure 2.1 CE marking

design dossier, which is a detailed technical file. The application must describe the design, manufacture, and performances of the product in question.

The notified body must examine the application and—if the product conforms to the relevant provisions of the directive—issues the application with an EC design-examination certificate.

Further requirements focus on the postmarketing surveillance and vigilance system.

In the case of devices in Classes IIa and IIb, all the requirements of Class III apply with the exception of the examination of the *design dossier*. The notified body, however, shall assess the technical documentation for representative samples.

Annexes III–VII give alternative options for the declaration of conformity.

Annex VIII regulates statements concerning devices for special purposes.

Clinical evaluation is the topic of Annex X, which groups requirements into two sets, general provisions and clinical investigations.

Annex XI defines criteria to be met for the designation of notified bodies.

Annex XII specifies the CE conformity marking, which consists of the initials CE placed at a precise distance and shaped proportionally as shown in Figure 2.1. If the marking is reduced or enlarged, the proportions given must be respected. Such specifications are definitively relevant to distinguish the original CE marking from others very similar.

2.3 The approval process for software as a medical device

Stand-alone software, after the amendment of Directive 2007/47/EC and under certain conditions, is considered to be an "active medical device" [11]. As a consequence, a manufacturer must get a certificate and affix the CE marking before putting its product on the European Market. The process for the certification of a medical device software consists of the following steps (Figure 2.2).

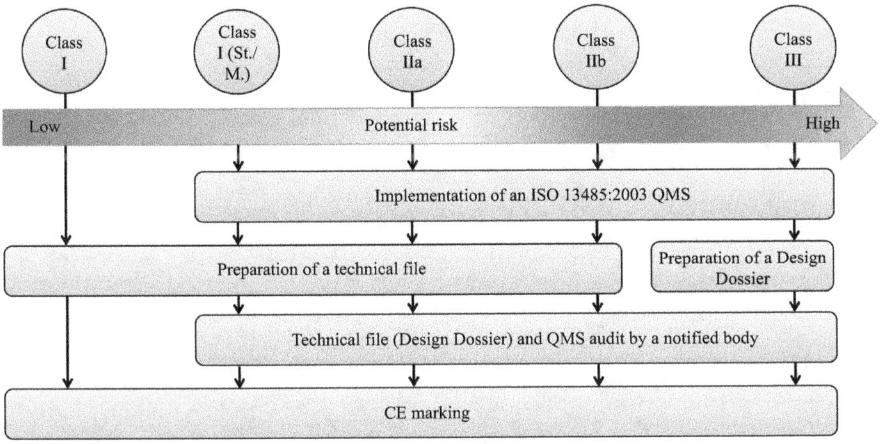

Figure 2.2 Approval process in Europe

2.3.1 Qualification

Qualification aims at establishing if the stand-alone software is to be treated as an (active) medical device (software as medical device), or not. It must have a medical purpose to be qualified as a medical device. The intended purpose, as described by the manufacturer of the product, is the unique element relevant for the qualification of any device.

Stand-alone software, which is intended by the manufacturer to be an accessory to a medical device (IVDMD), falls under the scope of Directive 93/42/EEC (Directive 98/79/EC), although it does not directly meet the definition of medical device (IVDMD).

Even if a product does not fall within the definition of medical device, it may be subject to other community and/or national legislations.

Software can be used for a large variety of medical purposes. Stand-alone software can (1) directly control equipment, (2) provide immediate decision triggering information, or (3) support in some way health-care professionals. Not all stand-alone medical software can be qualified as a medical device. Stand-alone software may run on different operating systems or in different virtual environments. These operating systems or environments do not impact on the qualification criteria. Stand-alone software might also be an accessory to a medical device. The risk related to a failure of the software used within healthcare is in itself not a criterion for its qualification as a medical device. It is, therefore, necessary to clarify some criteria for the qualification of stand-alone software as a medical device. The decision diagram shown in Figure 2.3 gives some guidance regarding the necessary steps to qualify stand-alone software as a medical device, as detailed in the EU "medical devices: guidance document—qualification and classification of stand-alone software" [11].

Step 1. If the product is a software according to the definition of this document, then it may be a medical device; if the product is not a software according to the definition of this document, then it is not covered by the medical device directives.

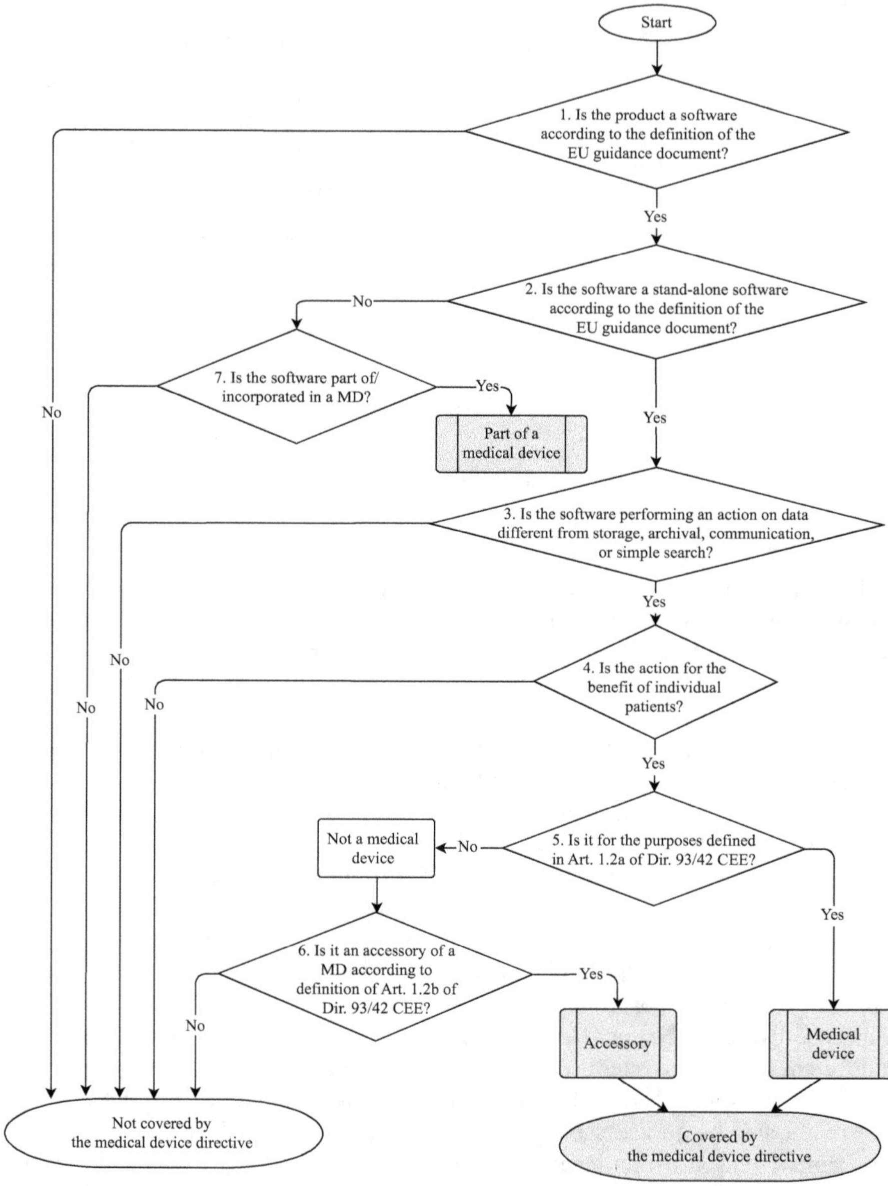

Figure 2.3 Decision diagram for the qualification of software as medical device

Step 2. If the software is incorporated in a medical device rather than stand-alone software, it must be considered as part of that medical device in the regulatory process of that device. If it is stand-alone software, proceed to decision Step 3.

Step 3. If the software does not perform an action on data, or performs an action limited to storage, archiving, communication, "simple searching" or lossless compression (i.e., using a compression procedure that allows the exact reconstruction of the original data), it is not a medical device. Altering the representation of data for embellishment purposes does not make the software a medical device. In other cases, including where the software alters the representation of data for a medical purpose, it could be a medical device.

"Simple searching" refers to the retrieval of records by matching record metadata against record search criteria, e.g., library functions. Simple searching does not include software which provides interpretative search results, e.g., to identify medical findings in health records or on medical images.

Software which is intended to create or modify medical information might be qualified as a medical device. If such alterations are made to facilitate the perceptual and/or interpretative tasks performed by health-care professionals when reviewing medical information (e.g., when searching the image for findings that support a clinical hypothesis as to the diagnosis or evolution of therapy), the software could be a medical device.

Note: The display of images usually involves alterations to the representation because techniques are used such as contrast stretching, edge enhancement, grayscale manipulation, smoothing, sharpening, zooming, and resizing. Alterations may include reconstruction, lossy compression, filtering, pattern recognition, modeling, interpolation, transformation, classification (e.g., scoring of tumors against specific criteria), segmentation, registration (e.g., mapping a data set to a model or atlas or to another data set, e.g., registering an Magnetic Resonance Imaging (MRI) image on a Computed Tomography (CT) image), calculations, quantification, qualification (e.g., comparison of data against references), rendering, visualization, interpretation, etc.

Step 4. An example of software for the benefit of individual patients is software intended to be used for the evaluation of patient data to support or influence the medical care provided to that patient. Examples of software which are not considered as being for the benefit of individual patients are those which aggregate population data, provide generic diagnostic or treatment pathways, scientific literature, medical atlases, models, and templates as well as software for epidemiological studies or registers.

Step 5. If the manufacturer specifically intends the software to be used for any of the purposes listed in Article 1(2)a of Directive 93/42/EEC, then the software shall be qualified as a medical device. However, if only a nonmedical purpose is intended by the manufacturer, such as invoicing or staff planning, it is not a medical device. Note: A task such as e-mailing, web or voice messaging, data parsing, word processing, and back-up is by itself not considered as being a medical purpose, according to Directive 93/42/EEC.

Step 6. If the software is an accessory to a medical device, it is not a medical device, but it falls under Directive 93/42/EEC.

The legal definition of "putting into service" requires that a device is made available to the final user/operator as being ready for use on the Community market. Software made available to the user over the internet (directly or via download) or via in vitro diagnostic commercial services, which is qualified as a medical device, is subject to the medical devices directives.

2.3.2 Classification

Once stand-alone software has been qualified as a medical device, it must be classified to determine its level of concern. Rules 9, 10, 11, and 12 of Annex IX of the directive may apply.

In the case that the software drives a medical device or influences the use of a device, it falls automatically into the same class as the device it drives.

In the case that the software acts as a stand-alone system, it is subject to the classification rules defined by the directive.

The guidance document [11] helps in moving within the jungle of the classification rules. It identifies the following groups of software systems.

2.3.2.1 Software as active therapeutic medical devices

All active therapeutic devices intended to administer or exchange energy are in Class IIa unless their characteristics are such that they may administer or exchange energy to or from the human body in a potentially hazardous way, taking account of the nature, the density, and site of application of the energy. In this case, they are in Class IIb (Rule 9 of Annex IX).

All active devices intended to control or monitor the performance of active therapeutic devices in Class IIb, or intended directly to influence the performance of such devices are, in Class IIb according to the implementation of Rule 2.3. Examples of Class IIb devices are radiotherapy planning systems used to calculate the dose of ionizing radiation to be administered to the patient and insulin dosage planning stand-alone software.

2.3.2.2 Software intended for diagnosis or therapy

According to Rule 10 of Annex IX, active devices intended for diagnosis are in Class IIa if they are intended to image in vivo the distribution of radio-pharmaceuticals (e.g., a clinical application of the registration of PET datasets on CT datasets for follow-up tumor treatment). Active devices are in Class IIa if they are intended to allow the direct diagnosis or monitoring of vital physiological processes. However, if they are specifically intended for the monitoring of vital physiological parameters where the nature of variations is such that it could result in immediate danger to the patient, for instance, variations in cardiac performance, respiration, or the activity of the Central Nervous System (CNS), they fall within Class IIb. For example, software for the presentation of the heart rate or other physiological parameters during routine checkups is in Class IIa, whereas software for the presentation of the heart rate or other physiological parameters for intensive care monitoring is in Class IIb.

Active devices intended to emit ionizing radiation and intended for diagnostic and therapeutic interventional radiology, including devices which control or monitor such devices, or which directly influence their performance, are in Class IIb.

Rule 11 of Annex IX states that all active devices intended to administer and/or remove medicines, body liquids or other substances to or from the body are in Class IIa, unless this is done in a manner that is potentially hazardous, taking account of the nature of the substances involved, of the part of the body concerned and of the mode of application. In such a case, they fall within Class IIb.

Stand-alone software which drives such a medical device or influences the use of such a device falls automatically into the same class as the device it drives (implementing Rule 2.3). Rule 12 of Annex IX states that all other active devices are in Class I. Class I stand-alone software can also come with a measuring function (e.g., orthopedic planning software to measure the interpedicular distance or sagittal diameter of the spinal canal).

2.3.2.3 IVDD and software in conjunction with in vitro diagnostic devices

Stand-alone software qualified as in vitro diagnostic medical devices should be regulated according to Directive 98/79/EC.

2.3.3 Selection of the Authorized Representative and notified body

The selection of an Authorized Representative is mandatory in any case where the manufacturer has no legal representation within the EU. Directive 2007/47/EC establishes that a non-EU manufacturer must designate a single EU/EC European Authorized Representative.

It is not mandatory to choose a notified body immediately, but it may be useful in case the manufacturer needs some assistance in the development process.

2.3.4 Implementation of a quality management system

The implementation of a Quality Management System (QMS) is mandatory for every medical device except those in Class I, as specified in Annexes II and V of the directive.

ISO 13485, which will be presented later in this book, is a harmonized standard. This means that a manufacturer who shows compliance with the European harmonized version of ISO 13485 automatically fulfills the QMS requirement posed by the directive.

2.3.5 Documenting software as a medical device

A medical device is generally required to be documented by means of a technical file, which is a comprehensive collection of data and documentation, to demonstrate compliance with the directive. For medical devices that fall within Class III, a more detailed document is needed—namely a Design Dossier—to support the examination of the design of the product.

In accordance with Recommendation NB-MED/2.5.1 [12], the technical documentation should include: Product Description, Technical Requirements, Design, Administrative Details (e.g., a declaration of conformity).

The amount of technical details increases with the level of concern for the device. In the case of a technical file, a high-level system architecture is sufficient. If the device falls within Class III and needs a Design Dossier, low-level architectural details must be provided.

A Technical File (Design Dossier) must be given to the notified body for the assessment.

2.3.6 Auditing by the notified body

Class I nonsterile and nonmeasuring devices do not need to be audited by the notified body. In all other cases (Classes IIa, IIb, and III devices and devices in List A or List B of Annex II) the device must be audited. In particular, their Technical File (or Design Dossier) along with their manufacturer's QMS (when required) will be examined.

2.3.7 Display of the CE marking

In the case where the audit with the notified body concludes positively, the manufacturer will receive the CE marking to affix.

Chapter 3
FDA title 21 of US CFR

3.1 The role of the Food and Drug Administration

In accordance with its mission, "The Food and Drug Administration (FDA) in the USA is responsible for protecting public health by assuring the safety, efficacy and security of human and veterinary drugs, biological products, medical devices, food supply, cosmetics, and products that emit radiation."

The FDA is also responsible for advancing public health by helping to speed up the progress of innovations that make medicines more effective, safer, and more affordable and by helping the public to obtain the accurate, science-based information they need to use medicines and foods to maintain and improve their health. The FDA also has responsibility for regulating the manufacturing, marketing, and distribution of tobacco products to protect the public health and to reduce tobacco use by minors.

Finally, the FDA plays a significant role in the national counter-terrorism capability. It fulfills this responsibility by ensuring the security of the food supply and by fostering the development of medical products to respond to deliberate and naturally emerging public health threats' [13].

As explained on its web site, the FDA has the legal authority to regulate both medical devices and electronic radiation-emitting products accordingly to the Federal Food Drug & Cosmetic Act (FD&C Act). The FD&C Act contains regulatory requirements that establish the FDA's level of control over these products. To fulfill the provisions of the FD&C Act that apply to medical devices and radiation-emitting products, the FDA develops, publishes, and implements regulations. It regulates a broad range of medical devices, including complicated, high-risk medical devices, like artificial hearts, and relatively simple, low-risk devices, like tongue depressors, as well as devices that fall somewhere in between, like sutures. It has the authority to regulate medical devices before and after they reach the marketplace, resulting their classification into one of the following groups:

FDA-listed medical devices. A medical device is FDA listed if the firm that manufactures or distributes has completed an online listing for the device through the FDA Unified Registration and Listing System (FURLS). (While manufacturers are the entities that typically list medical devices, they are not the only entities responsible for doing so.)

510(k) exempt medical devices. Medical devices that do not require a FDA review before the devices are marketed are considered "510(k) exempt." These medical devices are mostly low-risk Class I devices and some Class II devices that have been determined not to require a 510(k) (named from a section in the FD&C Act) to provide a reasonable assurance of safety and effectiveness.

These devices are exempt from complying with any premarket notification requirements subject to the limitations on exemptions; however, they are not exempt from certain general controls. For example, 510(k) exempt devices must (1) be suitable for their intended use; (2) be adequately packaged and properly labeled; (3) have establishment registration and device listing forms on file with the FDA; and (4) be manufactured under a quality system (with the exception of a small number of class I devices that are subject only to compliance files and general record-keeping requirements).

Cleared medical devices. These medical devices are ones that the FDA has determined to be substantially equivalent to another legally marketed device. A premarket notification, referred to as a 510(k), must be submitted to the FDA for clearance. A 510(k) is most often submitted by the medical device manufacturer.

Approved medical devices. Approved medical devices are those devices for which the FDA has approved a premarket approval (PMA) prior to marketing. This approval process is generally reserved for high-risk medical devices and involves a more rigorous premarket review than the 510(k) pathway.

3.2 Content of the Codes of Federal Regulation 21 CFR

The Federal FD&C Act and subsequent amending statutes are codified in Title 21 Chapter 9 of the United States Code. The Code of Federal Regulations (CFR) is a codification of the general and permanent rules that have been published in the Federal Register by the executive departments and agencies of the Federal Government. It is divided into 50 titles that represent broad areas subject to federal regulation. Most of the FDA's medical device and radiation-emitting product regulations are in Title 21 CFR Parts 800-1299, as shown in Figure 3.1. These final regulations codified in the CFR cover various aspects of the design, clinical evaluation, manufacturing, packaging, labeling, and postmarket surveillance of medical devices. In addition, the regulations address standards and product reports that apply to radiation-emitting products. Most of the relevant sections of Title 21 in respect of medical devices are listed in Table 3.1.

As an example, the following is the content of Section **820.30—Design controls**:

(a) *General.*

1. Each manufacturer of any class III or class II device, and the class I devices listed in paragraph (a)(2) of this section, shall establish and maintain procedures to control the design of the device in order to ensure that specified design requirements are met.
2. The following class I devices are subject to design controls:
   (i) Devices automated with computer software and
   (ii) The devices listed in the following Sections: 868.6810 (Catheter, Tracheobronchial Suction), 878.4460 (Glove, Surgeon's), 880.6760 (Restraint, Protective), 892.5650 (System, Applicator, Radionuclide, Manual), and 892.5740 (Source, Radionuclide Teletherapy).

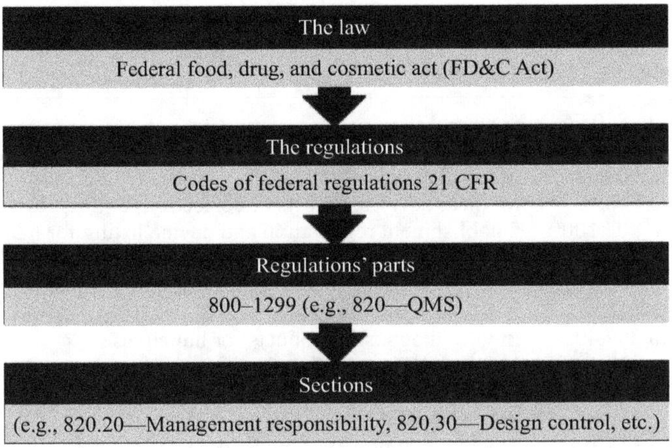

Figure 3.1 Structure of the regulations

(b) *Design and development planning*. Each manufacturer shall establish and maintain plans that describe or reference the design and development activities and define responsibility for implementation. The plans shall identify and describe the interfaces with different groups or activities that provide, or result in, input to the design and development process. The plans shall be reviewed, updated, and approved as design and development evolves.

(c) *Design input*. Each manufacturer shall establish and maintain procedures to ensure that the design requirements relating to a device are appropriate and address the intended use of the device, including the needs of the user and patient. The procedures shall include a mechanism for addressing incomplete, ambiguous, or conflicting requirements. The design input requirements shall be documented and shall be reviewed and approved by a designated individual(s). The approval, including the date and signature of the individual(s) approving the requirements, shall be documented.

(d) *Design output*. Each manufacturer shall establish and maintain procedures for defining and documenting design output in terms that allow an adequate evaluation of conformance to design input requirements. Design output procedures shall contain or make reference to acceptance criteria and shall ensure that those design outputs that are essential for the proper functioning of the device are identified. Design output shall be documented, reviewed, and approved before release. The approval, including the date and signature of the individual(s) approving the output, shall be documented.

(e) *Design review*. Each manufacturer shall establish and maintain procedures to ensure that formal documented reviews of the design results are planned and conducted at appropriate stages of the device's design development. The procedures shall ensure that participants at each design review include representatives of all functions concerned with the design stage being reviewed and an individual(s) who does not have direct responsibility for the design stage being reviewed, as well as any specialists needed. The results of a design review,

Table 3.1 Sections of the Codes of Federal Regulations 21 CFR

Part	Sections	Title
800	800.10 to 800.55	General
801	801.1 to 801.437	Labeling
803	803.1 to 803.58	Medical device reporting
806	806.1 to 806.40	Medical devices; reports of corrections and removals
807	807.3 to 807.100	Establishment registration and device listing for manufacturers and initial importers of devices
808	808.1 to 808.101	Exemptions from federal preemption of state and local medical device requirements
809	809.3 to 809.40	In vitro diagnostic products for human use
810	810.1 to 810.18	Medical device recall authority
812	812.1 to 812.150	Investigational device exemptions
813		[RESERVED]
814	814.1 to 814.126	Premarket approval of medical devices
820	820.1 to 820.250	Quality system regulation
821	821.1 to 821.60	Medical device tracking requirements
822	822.1 to 822.38	Postmarket surveillance
830	830.3 to 830.360	Unique device identification
860	860.1 to 860.136	Medical device classification procedures
861	861.1 to 861.38	Procedures for performance standards development
862	862.1 to 862.3950	Clinical chemistry and clinical toxicology devices
864	864.1 to 864.9900	Hematology and pathology devices
866	866.1 to 866.6050	Immunology and microbiology devices
868	868.1 to 868.6885	Anesthesiology devices
870	870.1 to 870.5925	Cardiovascular devices
872	872.1 to 872.6890	Dental devices
874	874.1 to 874.5900	Ear, nose, and throat devices
876	876.1 to 876.5990	Gastroenterology–urology devices
878	878.1 to 878.5910	General and plastic surgery devices
880	880.1 to 880.6992	General hospital and personal use devices
882	882.1 to 882.5975	Neurological devices
884	884.1 to 884.6200	Obstetrical and gynecological devices
886	886.1 to 886.5933	Ophthalmic devices
888	888.1 to 888.5980	Orthopedic devices
890	890.1 to 890.5975	Physical medicine devices
892	892.1 to 892.6500	Radiology devices
895	895.1 to 895.101	Banned devices
898	898.11 to 898.14	Performance standard for electrode lead wires and patient cables

including identification of the design, the date, and the individual(s) performing the review, shall be documented in the Design History File (the DHF).

(f) *Design verification.* Each manufacturer shall establish and maintain procedures for verifying the device design. Design verification shall confirm that the design output meets the design input requirements. The results of the design verification, including identification of the design, method(s), the date, and the individual(s) performing the verification, shall be documented in the DHF.

(g) *Design validation.* Each manufacturer shall establish and maintain procedures for validating the device design. Design validation shall be performed under defined operating conditions on initial production units, lots, or batches, or their equivalents. Design validation shall ensure that devices conform to defined user needs and intended uses and shall include testing of production units under actual or simulated use conditions. Design validation shall include software validation and risk analysis, where appropriate. The results of the design validation, including identification of the design, method(s), the date, and the individual(s) performing the validation, shall be documented in the DHF.

(h) *Design transfer*. Each manufacturer shall establish and maintain procedures to ensure that the device design is correctly translated into production specifications.

(i) *Design changes*. Each manufacturer shall establish and maintain procedures for the identification, documentation, validation or where appropriate verification, review, and approval of design changes before their implementation.

(j) *Design history file*. Each manufacturer shall establish and maintain a DHF for each type of device. The DHF shall contain or reference the records necessary to demonstrate that the design was developed in accordance with the approved design plan and the requirements of this part.

3.3 The approval process for Software as a Medical Device

Over the years, the FDA has issued several guidance documents to address the problem of handling software as a medical device:

1. "Medical Device Data Systems, Medical Image Storage Devices, and Medical Image Communications Devices" [14];
2. "General Wellness: Policy for Low Risk Devices" [15];
3. "Guidance for the Content of Premarket Submissions for Software Contained in Medical Devices" [16];
4. "General Principles of Software Validation" [17];
5. "Guidance for Industry, FDA Reviewers and Compliance on Off-The-Shelf Software Use in Medical Devices" [18];
6. "Mobile Medical Apps" [19].

The approval process for Software as a Medical Device is organized, in accordance with the previous documents, in the following steps.

3.3.1 Qualification

Some of the previous guidance documents help in the qualification of a software system as a Medical Device; i.e., they define rules or report examples concerning the process of establishing whether or not a software system falls under the regulatory environment.

For example, within the "Medical Device Data Systems, Medical Image Storage Devices, and Medical Image Communications Devices" guidance document the FDA has determined that these types of device pose a low risk to the public. Consequently, the FDA does not intend to enforce compliance with the regulatory controls.

Another class of software systems that typically does not fall under the regulatory environment is that of low-risk general-wellness applications. The "General Wellness: Policy for Low Risk Devices" guidance document informs that the FDA does not intend to examine low-risk "general-wellness products." The guidance identifies general-wellness products as those meeting two requirements: (1) they are intended for only general-wellness use, as defined in the guidance and (2) they present a very low level of risk for the user's safety.

Examples of general-wellness products are exercise equipment, video games, software programs, and other products that are available from retail establishments, when consistent with the two factors listed above.

The decision diagram depicted in Figure 3.2 shows an algorithm for establishing whether or not a software product can be considered a general-wellness product. The questions to answer are as follows:

Q1: Does the product have an intended use that relates to maintaining or encouraging a general state of health or a healthy activity?

Q2: Does the product have an intended use that relates the role of healthy lifestyle with helping to reduce the risk or impact of certain chronic diseases or conditions?

Q3: Is the product low risk?

In any case in which Q1 and Q3 are true or Q2 and Q3 are true, the product is likely to be a general-wellness product within the scope of this guidance, but the factors and examples in the guidance should be reviewed to confirm the status of the product.

The "Mobile Medical App" guidance document clarifies that Mobile Apps may fall within the category of Medical Devices. In such a case, they are called Mobile Medical Apps, and manufacturers must meet the requirements associated with the corresponding Medical Device Class.

The guidance document provides a (not exhaustive) list of mobile medical apps, which are subject to regulatory oversight:

1. Mobile apps that are an extension of one or more medical devices by connecting to such device(s) for purposes of controlling the device(s) (e.g., mobile apps that control the delivery of insulin on an insulin pump) or for use in active patient monitoring or analyzing medical device data.
2. Mobile apps that transform the mobile platform into a regulated medical device by using attachments, display screens, or sensors or by including functionalities similar to those of currently regulated medical devices (e.g., a mobile app that uses the built-in accelerometer on a smartphone to collect motion information to monitor sleep apnea).
3. Mobile apps that become a regulated medical device (software) by performing patient-specific analysis and providing patient-specific diagnosis, or treatment recommendations. These types of mobile medical apps are similar to or perform the same function as those types of software devices that have been previously

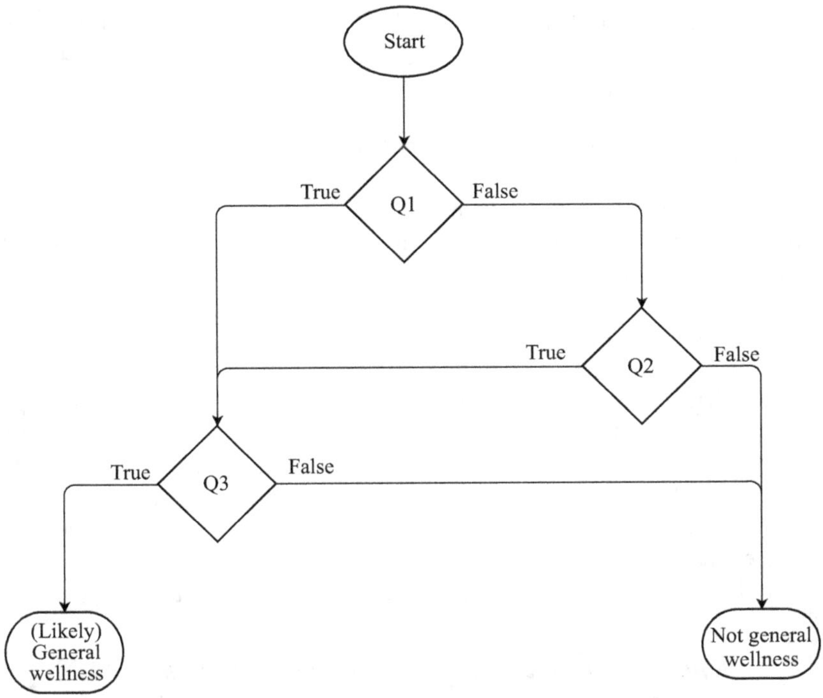

Figure 3.2 Decision diagram for general-wellness products

cleared or approved (e.g., apps that use patient-specific parameters and create a dosage plan for radiation therapy).

The guidance document also provides a list of mobile apps for which the FDA intends to exercise enforcement discretion; that is, the FDA does not intend to enforce any requirements under the FD&C Act. As an example, mobile apps that enable their users to get access to PHR systems, or EHR systems, fall within this category.

More generally, in order to qualify the software system as a Medical Device, it is possible to refer to existing databases available through the FDA web site. Otherwise, one must refer to the legal definition of Medical Device (i.e., "An instrument, apparatus, implement, machine, ..., or accessory which is: recognized in the official National Formulary, ..., intended for use in the diagnosis of disease ..., or intended to affect the structure or any function of the body of man or ..."): any software that meets such a legal definition is a device and must be realized in compliance with the requirements posed by the FD&C Act.

3.3.2 Classification

Title 21 of the CFR, rather than providing a set of general rules, defines three device categories: Class I (subject to general controls), Class II (subject to both general controls and special controls), and Class III (subject to premarket approval). Each category has a regulation number and assigns the risk classification accordingly.

Class I devices require general controls to provide a reasonable assurance of the safety and effectiveness of the device. Many devices falling in this category are exempt from any premarket notification and/or the Quality System Requirements, though they still have to comply with the other general controls.

Class II devices pose a higher level of risk to the patient and the general controls by themselves are insufficient to ensure the necessary degree of safety and effectiveness. Class II may include new devices for which information or special controls are available to reduce or mitigate the risk. Special controls may concern mandatory performance standards and postmarket surveillance.

Class III devices are those involving an unreasonable risk of illness or injury, or those for which there exists no evidence that the application of both general and special controls, or general controls only, is sufficient to provide confidence of the safety and effectiveness of the device. In other words, the fulfillment of the requirements for Class I and Class II devices are not sufficient to assure the required degree of safety and effectiveness. Class III includes devices which are life-supporting or life-sustaining. New devices that are not classified in Class I or II are automatically designated as Class III unless the manufacturer issues a request for reclassification.

3.3.3 Implementation of a quality management system

The quality system requirements are contained in Title 21 Part 820 of the CFR and reported in the following subsections.

3.3.3.1 Management responsibility

1. **Quality policy**—Management with executive responsibility shall establish its policy and objectives for, and commitment to, quality. Management with executive responsibility shall ensure that the quality policy is understood, implemented, and maintained at all levels of the organization.
2. **Organization**—Each manufacturer shall establish and maintain an adequate organizational structure to ensure that devices are designed and produced in accordance with the requirements of this part.
   - (i) *Responsibility and authority*—Each manufacturer shall establish the appropriate responsibility, authority, and interrelation of all personnel who manage, perform, and assess work affecting quality and provide the independence and authority necessary to perform these tasks.
   - (ii) *Resources*—Each manufacturer shall provide adequate resources, including the assignment of trained personnel, for management, performance of work, and assessment activities, including internal quality audits, to meet the requirements of this part.
   - (iii) *Management representative*—Management with executive responsibility shall appoint, and document such appointment of, a member of management who, irrespective of other responsibilities, shall have established authority over and responsibility for:

   (a) Ensuring that quality system requirements are effectively established and effectively maintained in accordance with this part and
   (b) Reporting on the performance of the quality system to management with executive responsibility for review.
3. **Management review**—Management with executive responsibility shall review the suitability and effectiveness of the quality system at defined intervals and with sufficient frequency according to established procedures to ensure that the quality system satisfies the requirements of this part and the manufacturer's established quality policy and objectives. The dates and results of quality system reviews shall be documented.
4. **Quality planning**—Each manufacturer shall establish a quality plan which defines the quality practices, resources, and activities relevant to devices that are designed and manufactured. The manufacturer shall establish how the requirements for quality will be met.
5. **Quality system procedures**—Each manufacturer shall establish quality system procedures and instructions. An outline of the structure of the documentation used in the quality system shall be established where appropriate.

3.3.3.2 Quality audit

Each manufacturer shall establish procedures for quality audits and conduct such audits to assure that the quality system is in compliance with the established quality system requirements and to determine the effectiveness of the quality system. Quality audits shall be conducted by individuals who do not have direct responsibility for the matters being audited. Corrective action(s), including a reaudit of deficient matters, shall be taken when necessary. A report of the results of each quality audit, and reaudit(s) where taken, shall be made and such reports shall be reviewed by management having responsibility for the matters audited. The dates and results of quality audits and reaudits shall be documented.

3.3.3.3 Personnel

1. **General**—Each manufacturer shall have sufficient personnel with the necessary education, background, training, and experience to assure that all activities required by this part are correctly performed.
2. **Training**—Each manufacturer shall establish procedures for identifying training needs and ensure that all personnel are trained to adequately perform their assigned responsibilities. Training shall be documented.
   (i) As part of their training, personnel shall be made aware of device defects which may occur from the improper performance of their specific jobs.
   (ii) Personnel who perform verification and validation activities shall be made aware of defects and errors that may be encountered as part of their job functions.

3.3.4 Documenting the Software as a Medical Device

The "Guidance for the Content of Premarket Submissions for Software Contained in Medical Devices" describes the requirements concerning the documentation that the

FDA recommends including in a premarket submission. The kind of documentation depends on the device's Level of Concern, which is intended as an estimate of the severity of the injuries that the device could cause to its patients (or any other kind of user) as a result of system failures, design flaws, or simply by employing the device for its intended use.

The Level of Concern is defined as Major (in any case in which a failure could directly or indirectly result in death or serious injury to the patient or any operator), Moderate (in any case in which a failure could directly or indirectly result in minor injury to the patient or any operator), or Minor (in the case a failure is unlikely to cause any injury to the patient or any operator). However, this Level of Concern assessment is not directly related to the device classification (Class I, II, or III).

In the following, some recommendations for the determination of the Level of Concern appropriate for a software device and some recommendations for the documentation that should be prepared for each Level of Concern are extracted from the guidance document.

The FDA recommends that the Level of Concern is determined and reported in the documentation before any mitigation of relevant hazards. It is also suggested that the method for the determination of the Level of Concern is described in the documentation.

It has also provided two lists of questions to answer in order to assess the level of concern as major or moderate. By exclusion, if the answer is "no" to all these questions, the Level of Concern is assessed as minor.

When the level of concern has been determined, the manufacturer can refer to Figure 3.3 to establish the kind of documentation to produce for the software system under construction.

3.3.5 FDA clearance and premarket approval

The FDA requires that all manufacturers register their facilities, list their devices with the agency, and follow general controls requirements. Many medical devices (including, of course, software as a medical device) involve only minimal risks and can be freely marketed after having completed the registration. Such a category of low-risk systems is deemed exempt from premarket review and manufacturers do not have to submit an application to the FDA before marketing. In contrast, most other software applications, falling within either the Moderate or Major Level of Concern assessment groups, must obtain the agency's permission before any commercialization. The FDA grants this permission when a manufacturer meets the regulatory premarket requirements and agrees to any necessary postmarket requirements which vary according to the risk that the device presents.

There are two routes that manufacturers can follow to place their medical devices on the market with the FDA's permission. The first requires that the manufacturer conducts clinical studies, submits a PMA application and provides evidence indicating with reasonable confidence that the device is safe and effective. The PMA route is generally used for brand new and high-risk devices. The second path involves the preparation and submission of a 510(k) notification that aims at

Software documentation	Level of concern		
	Minor	Moderate	Major
Level of concern	A statement indicating the level of concern and a description of the rationale for that level		
Software description	A summary overview of the features and software operating environment		
Device hazard analysis	Tabular description of identified hardware and software hazards, including severity assessment and mitigations		
Software requirements specification	Summary of functional requirements from SRS	The complete SRS documents	
Architecture design chart	No documentation is necessary in the submission	Detailed depiction of functional units and software modules. May include state diagram as well as flow charts	
Software design specification	No documentation is necessary in the submission	Software design specification document	
Traceability analysis	Traceability among requirements, specifications, identified hazards and mitigations, and verification and validation testing		
Software development environment description	No documentation is necessary in the submission	Summary of software life cycle development plan, including a summary of the configuration management and maintenance activities	Summary of software life cycle development plan. Annotated list of control documents generated during development process. Include the configuration management and maintenance plan documents
Verification and validation documentation	Software functional test plan, pass /fail criteria, and results	Description of V&V activities at the unit, integration, and system level. System level test protocol, including pass/fail criteria, and tests, results	Description of V&V activities at the unit, integration, and system level. Unit, integration, and system level test protocols, including pass/fail criteria, test report, summary, and tests results
Revision level history	Revision history log, including release version number and date		
Unresolved anomalies (bugs or defects)	No documentation is necessary in the submission	List of remaining software anomalies, annotated with an explanation of the impact on safety or effectiveness, including operator usage and human factors	

Figure 3.3 Documentation for the software based on the Level of Concern

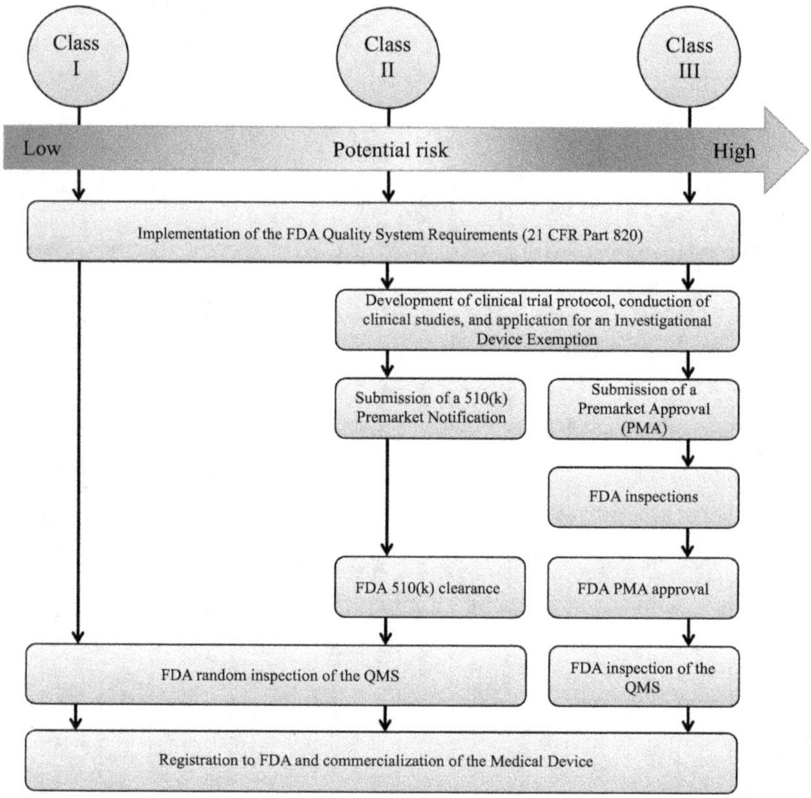

Figure 3.4 Approval process in the USA

demonstrating that the device under analysis is substantially equivalent to another one (a predicate device) already approved for the market. Both devices must have the same intended use and technological characteristics, as well as an equivalent performance. In the case of approval of this 510(k), the permission is called "Clearance."

The review performed by the FDA is substantially different in the case of PMA approvals from that relating to a 510(k). Indeed, a PMA review consists of assessing the degree of safety and efficacy of the device, which must provide a sufficient level of confidence. On the contrary, a 510(k) review aims at determining whether the device and its predicate are substantially equivalent (Figure 3.4).

In order to assess whether or not a medical device is compliant the FDA looks for evidence that a structured process was used to develop the device software. No specific development process model is prescribed but the use of a combination of software verification and validation activities is recommended. In particular, it is suggested that software verification and validation be conducted, including a variety of static and dynamic techniques, code inspections and document reviews. Moreover, the FDA requires that such software validation and verification activities are performed in compliance with the Quality System Regulation defined in 21 CFR Part 820.

Chapter 4
Regulations for other markets

4.1 Regulatory environment and approval process in Australia

The Therapeutic Goods Administration (TGA) controls medical devices in Australia according to criteria prescribed by the Therapeutic Goods (Medical Devices) Regulations 2002 [20].

The regulatory environment in Australia is quite similar to that in Europe. Indeed, if a device has the European CE Marking, the classification will probably be consistent and the approval process will be easier since the CE Marking is generally accepted by the TGA.

The Therapeutic Goods (Medical Devices) Regulations 2002 Schedule 2 defines the classification rules for medical devices, which are based on (1) the manufacturer's intended use, (2) the level of risk, and (3) the degree of invasiveness in the human body. Five classes of medical devices have been established: *Class I*, *Class IIa*, *Class IIb*, *Class III*, and *Active Implantable Medical Devices* (AIMD).

The classification determines the conformity assessment procedure that the manufacturer can choose to ensure conformity with the requirements introduced by the Australian regulations. The higher is the class of the device, the more stringent is the set of requirements they are required to undergo. The responsibility for the conformity assessment belongs to the manufacturer. For devices imported into Australia and falling within Class I, the manufacturer self-certifies and signs a declaration of conformity. For higher classes of devices, the TGA issues the certification after confirming that the conformity assessment procedure is correct and has been applied.

All registrable devices undergo a premarket evaluation before having access to the market. Devices cleared for the market are provided with an Australian Register of Therapeutic Goods (ARTG) number.

The approval process consists of the following main steps (Figure 4.1):

Step 1. Determining the classification through Schedule 2 of the Australian Therapeutic Goods (Medical Devices) Regulations 2002. In any case in which the device has already been certified for the European Market, the resulting class is likely to be the same. CE Marking certificates from European Notified Bodies are generally accepted by the TGA as part of any registration.

If the manufacturer has no local presence in Australia, an Australian Sponsor must be appointed. The Sponsor facilitates the device registration and acts as the

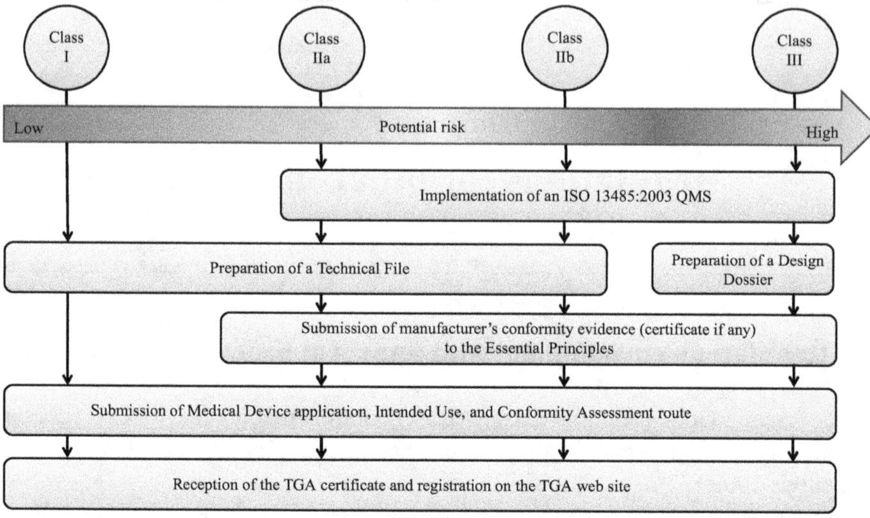

Figure 4.1 Approval process in Australia

liaison between the manufacturer and the TGA. The Sponsors name must appear on the device and labeling.

Step 2. Implementing a QMS. ISO 13485 is a recognized standard.

Step 3. Depending on the device classification, making the Technical File or Design Dossier (Class III) and Declaration of Conformity ready for submission.

Step 4. For all devices except Class I, the Sponsor submits the Manufacturer's Evidence (CE Marking certificate) for the TGA's review and acceptance.

Step 5. For all devices, the Sponsor submits the Medical Device application, along with the intended use and Conformity Assessment route.

Step 6. If approved by the TGA, the issuing of an ARTG number.

4.2 Regulatory environment and approval process in Brazil

In Brazil, the ANVISA (Agência Nacional de Vigilância Sanitária) is the authority that regulates medical devices, and Resolution RDC No. 185 of the October 22, 2001 [21] is the main regulation pertaining to medical devices.

The classification system, described in Annex II of RDC No. 185, is rules based and associates devices to four classes: *Class I*, *Class II*, *Class III*, or *Class IV*. The classification structure corresponds to that used in the European Union under Council Directive 93/42/EEC.

All medical devices distributed within Brazil must first be registered with ANVISA. The registration of most Class I and II devices involves a relatively simple process referred to as "Cadastro." A more rigorous process (called "Registro") applies

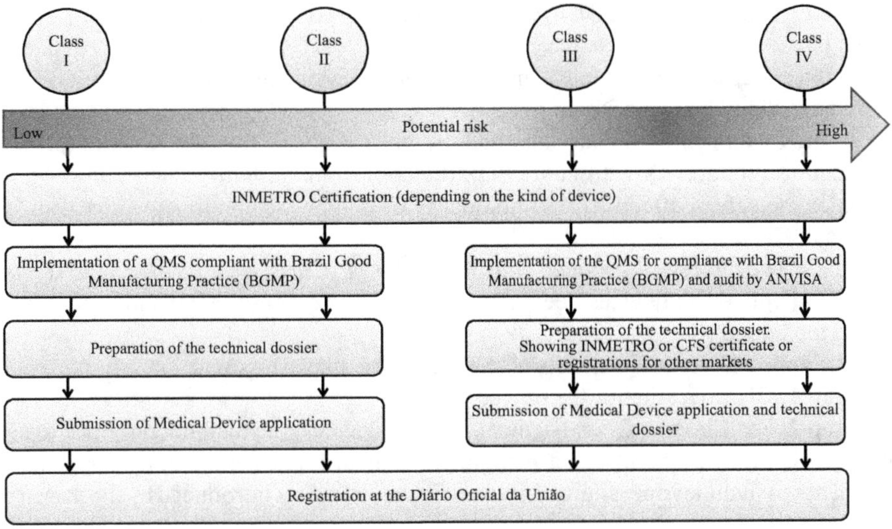

Figure 4.2 Approval process in Brazil

to some Class I and Class II medical devices, as well as all Class III and Class IV devices.

The approval and registration process consists of the following steps (Figure 4.2):

Step 1. Classification of the device using rules found in Annex II of Resolution RDC 185/2001. If the manufacturer has no legal representative in Brazil, a Brazil Registration Holder (BRH) must be appointed.

Step 2. Some devices require a certification from the Brazil's Institute of Metrology, Standardization and Industrial Quality (INMETRO), which is the body responsible for the implementation of measurement, safety, and quality standards for electrical and electronic products.

Step 3. Class I and II device manufacturers must comply with the Brazil Good Manufacturing Practice (BGMP) requirements. In the case of Class III or IV devices, the manufacturer's QMS is audited by ANVISA.

Step 4. Class I and II devices need Technical Dossiers. The technical documentation is maintained by the BRH for ANVISA on-site inspections. Manufacturers of Class III or IV devices should alternatively obtain a Certificate of Free Sale (CFS), or INMETRO certificate, or show certificates and registrations for at least two other markets.

Step 5. The BRH prepares and submits the application, which, in the case of Class III and IV devices, includes the technical dossier, to ANVISA.

Step 6. ANVISA assesses the registration application. On approval, it will publish the registration number in the Diário Oficial da União (DOU). Class I and II registrations do not expire. Class III and IV registrations have a limited period of validity up to 5 years.

4.3 Regulatory environment and approval process in Canada

Medical devices traded in Canada are subject to the Medical Device Regulation [22] of the Canadian Food and Drugs Act.

Approvals for trading are handled by the Medical Devices Bureau of the Therapeutic Products Directorate (TPD), the competent authority that monitors and evaluates the safety, effectiveness and quality of diagnostic and therapeutic medical devices in Canada.

Medical devices are grouped into four classes: *Class I*, *Class II*, *Class III*, or *Class IV*. Higher risk devices need a Medical Device License (MDL) in order to be sold in the country.

The approval process consists of the following steps (Figure 4.3):

Step 1. Classification of the device.

Step 2. For Class II, Class III, and Class IV devices, the manufacturer has to set up a QMS. ISO 13485:2003 is a recognized standard, but the quality management system has to include some additional specific requirements introduced by the national regulation.

Step 3. For Class II, Class III, and Class IV devices, the manufacturer's QMS is audited by the competent authority.

Step 4. Submission of an application along with all necessary documents: Medical Device Establishment License (MDEL) application for Class I devices, Canadian MDL application and ISO 13485 certificate for Class II devices and MDL application, ISO 13485 certificate and the Premarket Review Document for Class III and Class IV devices.

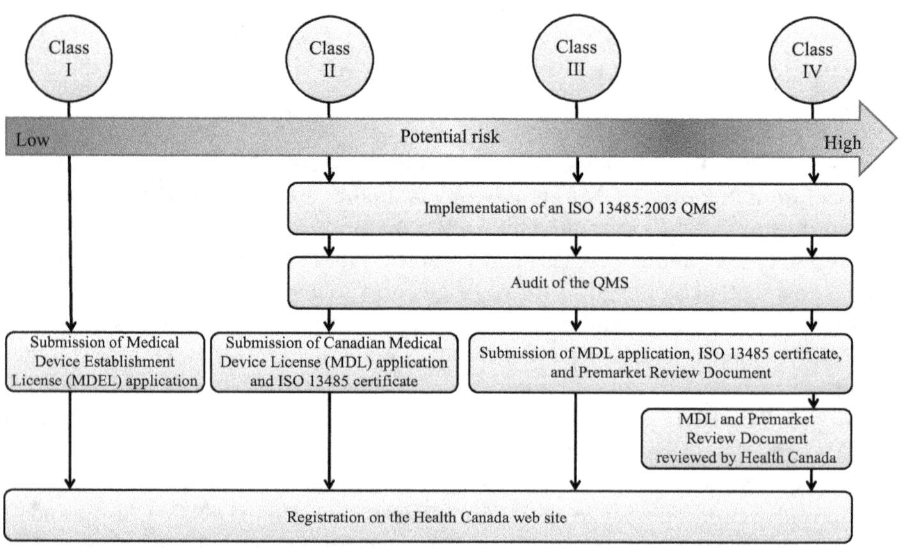

Figure 4.3 Approval process in Canada

Step 5. For Class IV devices, the submitted documents are reviewed by Health Canada.

Step 6. Registration at the Health Canada web site.

4.4 Regulatory environment and approval process in China

The China Food and Drug Administration (CFDA), previously known as the State Food and Drug Administration (SFDA), controls the market for medical devices in China.

The classification of medical devices in China follows the requirements of SFDA Order No. 15 [23]. It presents some relevant deviations from the classification schemes in other countries, such as the USA or the European Union. Medical Devices are grouped into three categories: *Class I*, *Class II*, and *Class III* .

For importers, the registration process requires the submission of a registration standard along with device samples for testing. Manufacturers of Class II and Class III medical devices are also required to demonstrate that the device has been approved by the country of origin with documents such as a CE certificate, 510(k) letter and PMA approval and compliance with ISO 13485 and may also be required to submit clinical data in support of their application. In addition to these requirements, all medical device manufacturers must also include product information in Chinese on all packaging and labeling, as detailed in SFDA Order No. 10.

The approval process consists of the following steps (Figure 4.4):

Step 1. Classification of the device using rules found in the CFDA Order No. 15. If the manufacturer has no legal representative in China, a China-based agent must be appointed.

Step 2. Manufacturers must demonstrate the implementation of a QMS compliant to ISO 13485 or US FDA QSR (21 CFR Part 820) requirements.

Step 3. For Class II and Class III, the device must be submitted to the CFDA.

Step 4. Execution of additional clinical trials and testing in China (Class I devices are exempt from this requirement).

Step 5. For Class I devices, an application for an Import Medical Device Registration Certificate (IMDRC) must be prepared. For Classes II and III, along with the IMDRC application, the Registration Standard Dossier must be submitted.

Step 6. In the case of approval, the CFDA releases the IMDRC.

4.5 Regulatory environment and approval process in Japan

The Pharmaceutical Affairs Law (PAL), revised on November 2014, provides the legal framework for the regulation of medical devices in Japan. The Pharmaceuticals and Medical Devices Agency (PMDA) was established in April 2004 in order to handle all consultation and review work and the postmarketing surveillance of medical devices.

In Japan, medical devices are classified as *Class I*, *Class II*, *Class III*, or *Class IV*.

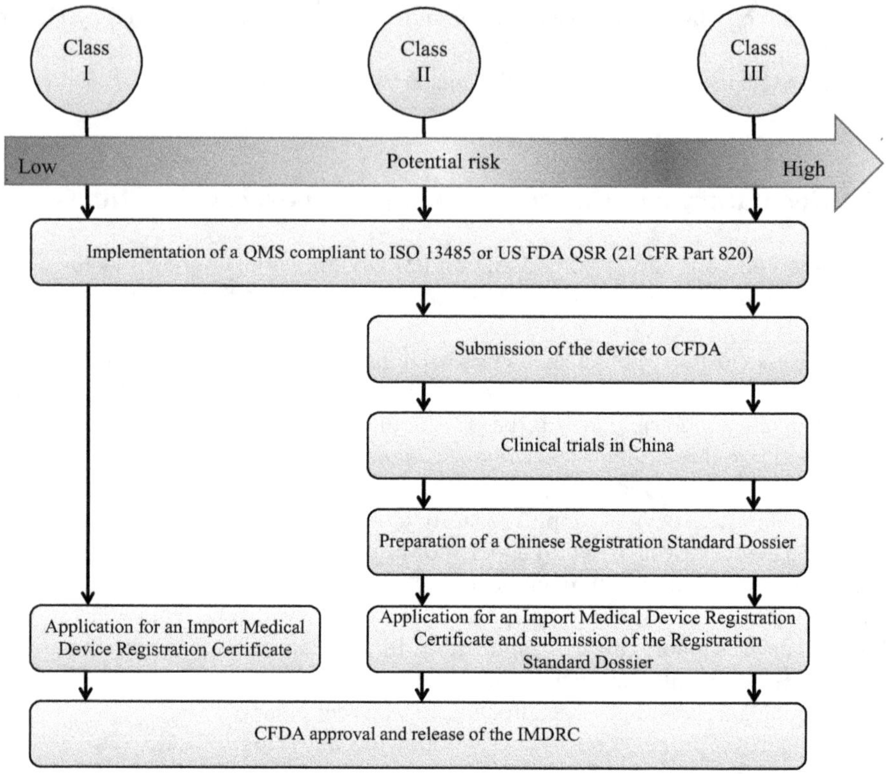

Figure 4.4 Approval process in China

The Marketing Authorization Holder (MAH) must register the device. The procedure depends on the class of device:

Premarket Submission (*Todokede*). To register and market a General Medical Device (a Class I device), the MAH only needs to file a Premarket Submission with the PMDA with no further assessment.

Premarket Certification (*Ninsho*) to register Class II devices which are described as "Specified Controlled Devices." In such a case, the MAH needs to file a Premarket Certification application with a registered certification body and obtain their certification.

Premarket Approval (*Shonin*) to register and market a Highly Controlled Medical Device (Classes II, III, and IV Medical Devices). The MAH needs to file a Premarket Approval Application with the PMDA and obtain approval from the Minister of Health, Labour and Welfare (MHLW). Class II devices that are not Specified Controlled Devices are also subject to Premarket Approval.

The approval process consists of the following steps (Figure 4.5):

Step 1. Classification of the device according to the Japanese Pharmaceutical and Medical Device Act (PMD Act) and Japanese Medical Devices Nomenclature

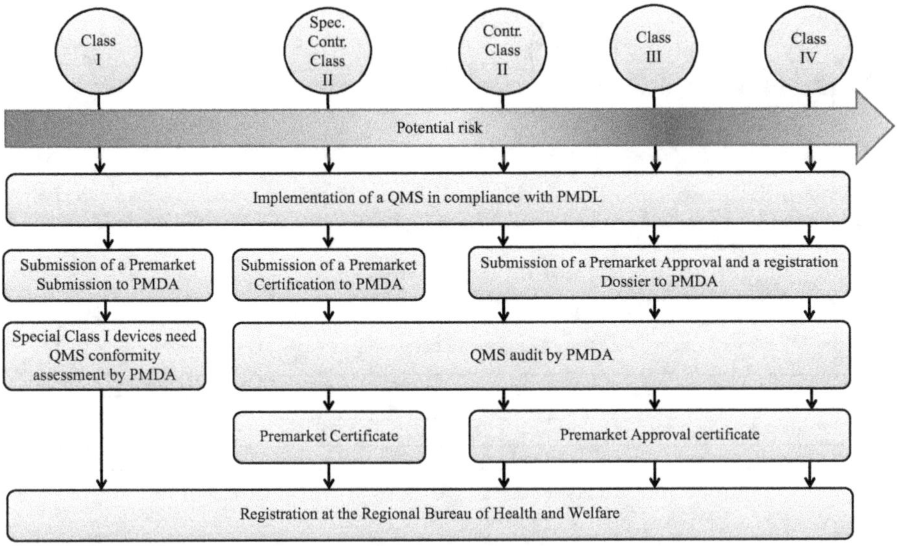

Figure 4.5 Approval process in Japan

(JMDN) codes. The manufactures must appoint a Marketing Authorization Holder (MAH or D-MAH) to manage the approval process.

Step 2. Implementation of a QMS that must be compliant with PMD requirements. For Controlled Class II, Class III, and Class IV devices, in addition to the prepared Premarket Approval application, a registration dossier must be submitted as a Summary Technical Document (STED).

Step 3. Submission of the registration application (*Todokede*, *Ninsho*, or *Shonin*).

Step 4. The QMS is either assessed or audited.

Step 5. Specified Controlled Class II devices will have a Premarket Certificate issued by the RCB. Controlled Class II, Highly Controlled Class III, and Highly Controlled Class IV devices will have a Premarket Approval certificate issued by the MHLW. Class I devices have no certificate issued.

Step 6. For all classes, an Import Notification to the Regional Bureau of Health and Welfare must be submitted.

4.6 Regulatory environment and approval process in Russia

The legal framework in the Russian Federation is established by the Federal Law on Fundamentals of Healthcare, which was implemented by the Government Regulation No. 1416 of December 27, 2012 [24]. The Federal Service for Surveillance in Healthcare (Roszdravnadzor) is the federal executive body responsible for the control and supervision of the health-care system.

The regulation defines classification and approval processes very similar to the ones active in the EU.

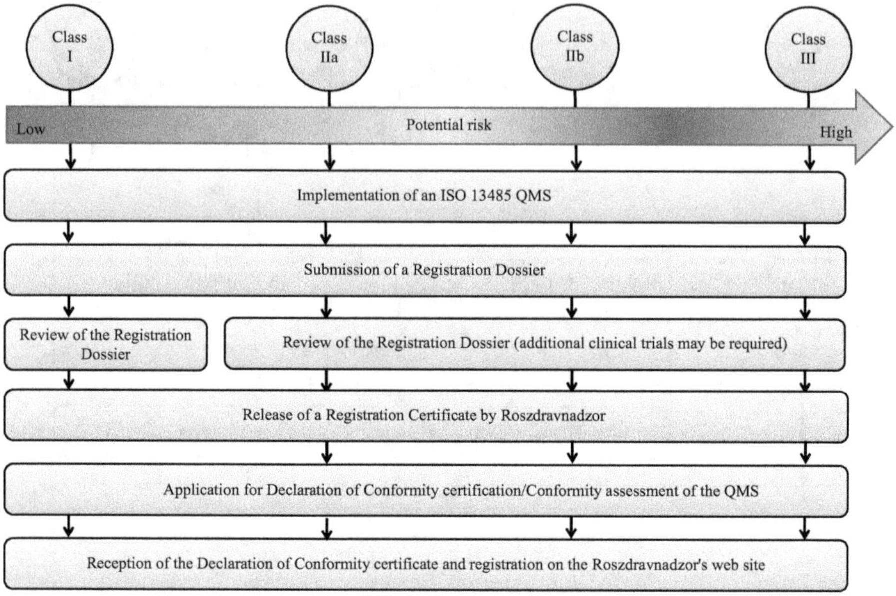

Figure 4.6 Approval process in Russia

The approval process consists of the following steps (Figure 4.6):

Step 1. Classification of the device using Roszdravnadzor Order No. 4. An Authorized Representative must be appointed to coordinate the registration operation in Russia.

Step 2. Implementation of an ISO 13485 QMS.

Step 3. Submission of the technical documentation (Registration Dossier) to an authorized testing center. Tests must be conducted in Russia even if equivalent verifications have already been performed in other countries.

Step 4. Roszdravnadzor reviews the Registration Dossier and, for devices falling within classes higher that I, the clinical trials.

Step 5. Roszdravnadzor releases a Registration Certificate.

Step 6. Application for a Declaration of Conformity certification or Conformity assessment of the QMS depending on the device class. For Class I devices, the Expertise Center reviews the dossier. For all other classes, Roszdravnadzor, along with their Expertise Center, will review the Registration Dossier and determine if additional clinical data, trials, or testing are needed.

Step 6. In the case of approval, Roszdravnadzor issues a Registration Certificate and adds the Medical Device to the database on its web site.

Part III

Standards

Chapter 5

ISO 13485: medical devices—quality management systems—requirements for regulatory purposes

5.1 Introduction

ISO 13485 (medical devices—quality management systems—requirements for regulatory purposes) is an international standard that presents the requirements for a quality management system specific for the realization of medical devices, including software systems with medical purposes.

Any organization adopting ISO 13485 can use it throughout the entire life cycle of the medical device, from its initial conception to post production and market surveillance. The standard can also be used by other parties, such as certification bodies and competent authorities, to help them with their certification processes and audit activities. More specifically, ISO 13485 applies to medical device suppliers and service providers of all kind. It also applies to any company or institution participating in one or more stages of the product life cycle. In addition, it applies to groups that design, develop, produce, store, distribute, or install medical devices and to those that design, develop, or provide activities related to medical devices. It may also apply to external suppliers that provide services, products, or facilities to these organizations. Regulatory bodies often require that organizations define the roles they play in the medical device supply chain. As a consequence, ISO 13485 expects the applying organization to specify the roles it plays (manufacturer, distributor, importer, developer, or designer of medical devices) and the regulatory requirements that apply. The organization should then build such requirements into its Quality Management System (QMS).

The adoption of ISO 13485 relates to the development of a QMS for medical devices. The use of this system is intended to (1) assess the organization's capability to provide medical device products that meet the requirements presented by the regulations; (2) demonstrate that the organization is able of providing medical device services and products that meet customer requirements; and (3) enable the organization to obtain the certification. In fact, once the organization has established a QMS that complies with the ISO 13485 standard, it can ask a certification body to audit its QMS and get an official certificate.

ISO 13485 does not require all organizations to structure their QMS or their documentation in the same way. On the contrary, this will depend on the organization's structure, complexity, needs of compliance, organizational processes, activities, products and services traded, risks, threats, and opportunities. Consequently, the structure and contents of a QMS vary case by case.

However, the hierarchy of documentation of the expected QMS is usually divided into four tiers (Figure 5.1):

1. the Quality Policy and Quality Manual
2. the Procedures
3. the Work Instructions
4. the Quality Records

The Quality Manual is the top level of the documentation stack. The purpose of this level is to declare the quality policies and objectives of the company, its organization and scope, any exclusions from ISO 13485 clauses, its quality processes, and quality procedures. Such policies are similar to mission statements and may influence management decisions.

The organization's procedures are the second level of the documentation and describe "who" does "what" and "when." Typically, they are referred to, in the quality system, as a "written procedure" or a "documented procedure." Activities might be listed along with the functional titles or positions responsible for the procedure. These procedures can be text based, but many use a process map, or a diagram, to communicate the information.

The work instructions provide a very detailed description on "how" to accomplish a specific task. They are very specific to an organization. Additional documentation, such as for example user and technical manuals, support notes, and manufacturing notes, may be employed to create detailed work instructions.

The bottom level of documentation regards forms adopted to create checklists, surveys, records, or other documents useful for the creation of the product. The quality records provide the foundation of the quality system. As defined by the International Standard on Records Management ISO 15489 [25], records are "information created, received and maintained as evidence and information by an organization or person, in pursuance of legal obligations or in the transaction of business." Quality records comprise all forms or controlled documents necessary for the quality system to operate. Records are a critical output of any work instruction or procedure. They form the basis of the process communications, audit material, and process improvement action.

For the reminder of the chapter, we will refer to ISO 13485:2016, which is the latest available version of the standard.

5.2 Contents

ISO 13485 specifies the requirements for a QMS that can be used by an organization involved in one or more stages of the life cycles of a medical device, including its design and development, production, storage and distribution, installation and servicing, and final decommissioning and disposal.

Figure 5.1 The documentation hierarchy

In ISO 13485, the following terms or phrases are used: a requirement is qualified as "As appropriate" if it is supposed to be appropriate unless the organization can justify otherwise. The term "Risk" is used in relation to safety or performance requirements of the medical device. A requirement is required to be "Documented" when it is also required to be established, implemented, and maintained. The term "Product" may also refer to a service product.

ISO 13485 uses the following verbs:

- *Shall* indicates a requirement
- *Should* indicates a recommendation
- *May* indicates a permission
- *Can* indicates a possibility or a capability

Concerning the requirements, the core sections of the standard are presented in Figure 5.2.

5.2.1 The Quality Management System

This section presents some general requirements (clause 4.1) and some documentation requirements for the QMS.

The general requirements are the foundation of why and how an organization establishes, implements, and maintains a successful QMS.

The organization "shall" document the QMS and maintain its effectiveness. It is also required to establish, implement, and maintain the procedures required by ISO 13485 or any applicable regulatory requirements. The roles (e.g., manufacturer, authorized representative, importer, or distributor) undertaken under the applicable regulatory requirements must also be documented. The organization is also supposed to determine the processes needed for the QMS, to apply a risk-management framework to the control of such processes, and to determine the interdependencies of these processes.

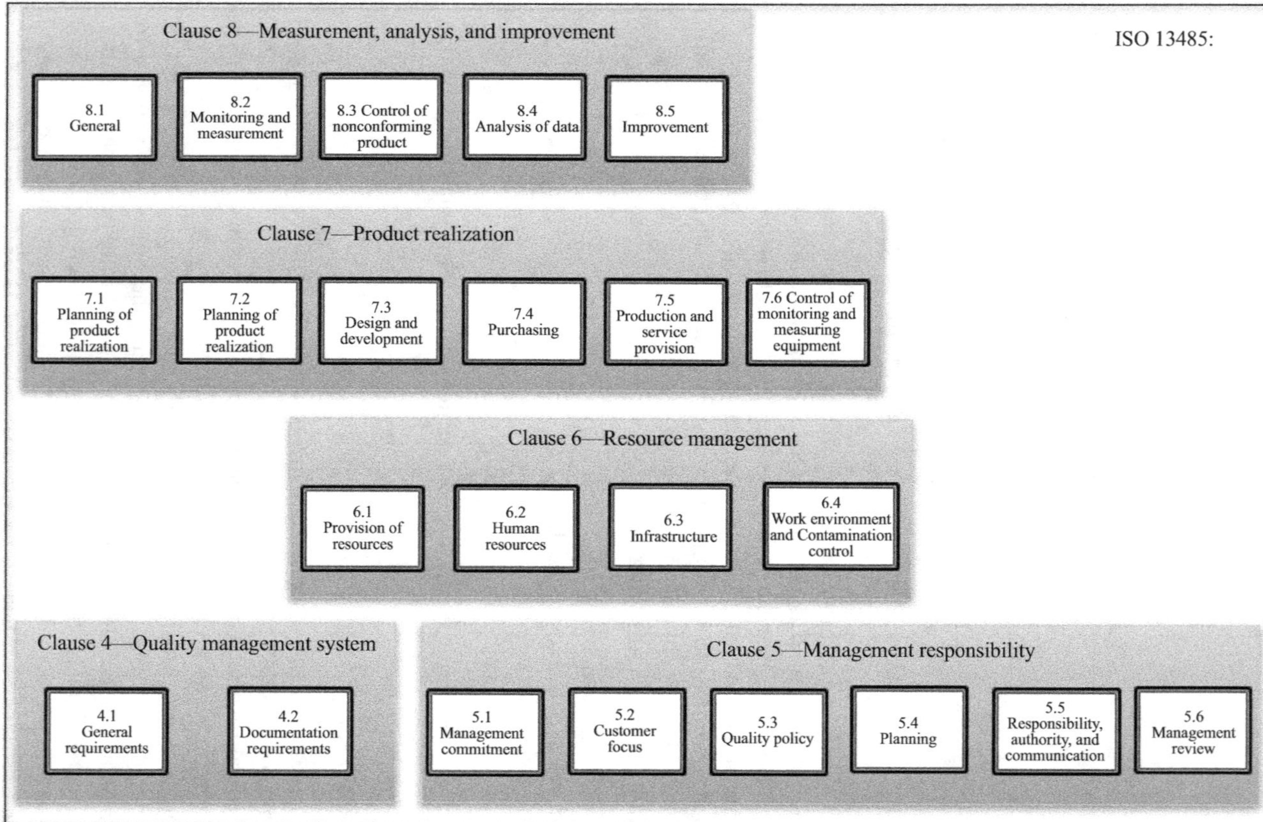

Figure 5.2 The structure of the ISO 13485:2016 Standard

The evaluation of the effectiveness of the operation and control of these processes is the responsibility of the organization. Adequate resources and tools (e.g., metrics) should be ensured to achieve this goal. Any change to these processes must be evaluated in terms of its impact on the QMS and the organization's products.

When the organization decides to outsource any process that affects product conformity to the requirements, it must monitor and ensure control over such processes. The organization must retain responsibility for conformity to the standard and to any applicable regulatory requirements for outsourced processes.

The organization must document all the procedures for the validation of the application of software systems adopted in the QMS. Such software applications must be validated before the initial use and, as appropriate, after changes to such software or its application. The approach and activities associated with software validation and revalidation must be proportionate to the risk associated with its use.

The QMS documentation shall include: documented statements of a quality policy and quality objectives; a quality manual; documented procedures and records required by the standard; documents, including records, determined by the organization to be necessary to ensure the effective planning, operation, and control of its processes; and other documentation specified by any applicable regulatory requirements.

The QMS documentation must include (1) documented statements of the quality policy and objectives; (2) a quality manual; (3) documented procedures, instructions, and records required by the standard; (4) documents identified by the organization to be necessary to ensure the effective planning, operation, and control of its processes; and (5) all other documentation specified by any applicable regulatory requirements.

The quality manual must include (1) the scope of the QMS, including details and a justification for any exclusion or nonapplicability; (2) the documented procedures for the QMS; and (3) a description of the interaction between the processes of the QMS.

For each type of Medical Device, the Standard requires that the organization establishes and maintains documents either containing or referencing records generated to demonstrate conformity to the requirement and compliance with any applicable regulatory requirements.

All documents and records required by the QMS must be controlled. A documented procedure must be realized to define the controls needed to review, approve, and update documents.

5.2.2 Management responsibility

Section 5 defines the requirement of the top management to ensure that managers actively participate in the QMS.

Clause 5.1—Management commitment—requires that the top management not only develops and implements a QMS, but also maintains its effectiveness.

The purpose of clause 5.2—Customer focus—is to ensure that customer requirements are determined and met.

A quality policy (clause 5.3) must be established. The quality policy must (1) be applicable to the purpose of the organization; (2) provide a framework for establishing and reviewing quality objectives; and (3) be communicated and understood within the organization.

Clause 5.4—Planning—covers quality objectives, which must be established at relevant functions and must be measurable and consistent with the quality policy. The responsibility, authority, and communication must be clearly defined and documented such as in an organizational chart and a management representative must be appointed and must be responsible for the effectiveness of the QMS. The top management is also responsible for internal communication regarding the effectiveness of the quality management system. There is also the requirement to hold management review meetings. The top management shall review the QMS at documented planned intervals to ensure its continuing suitability. The review must also include assessing any opportunities for improvement of or potential need for change to the QMS.

5.2.3 Resource management

Section 6 starts with clause 6.1—Provision of resources—that requires that the organization provides the resources needed to: (1) implement the QMS and maintain its effectiveness and (2) meet any applicable regulatory and customer requirements.

Concerning human resources (clause 6.2), the organization shall document the process for establishing competence, which means that the organization needs to determine the competence of personnel performing work affecting quality. It is necessary to provide training to maintain such competence, evaluate the effectiveness of that training, to ensure that the stuff are aware of how they contribute to customer satisfaction as well as the achievement of the quality objectives and to maintain records of education, training, skills, and experience.

Clause 6.3—Infrastructure—establishes that the organization must document the requirements for the work environment necessary to achieve product conformity, which includes how to prevent mix-ups and ensure an orderly handling of the product.

"Work environment and contamination control" is the clause (6.4) that requires that the organization document the requirements related to the work environment. Contamination control requires that the organization plans and documents arrangements to control contaminated products in order to prevent any contamination of the work environment.

5.2.4 Product realization

Section 7 is about product realization, namely how a company realizes the product and how it is delivered to the customer.

The manufacturer is required to develop product realization processes and to plan how products are realized by meeting quality objectives, identifying product realization requirements, and establishing arrangements for communicating with customers and regulators.

This section also concerns purchasing, production, and service provision. Purchase procedures must be prepared, in accordance with which the organization must control the selection of suppliers, monitor supplier performance, plan product purchases, and verify purchased products. In addition to this, the manufacturer must

(1) plan, monitor, and control Medical Device production and service provision; (2) define requirements for cleanliness or contamination controls; (3) specify product installation and verification requirements; (4) develop servicing procedures and reference materials; (5) validate processes used for production and service provision; and (6) facilitate product traceability.

It is also mandatory to determine the monitoring and measuring requirements, to establish monitoring and measurement procedures, to select suitable monitoring and measuring equipment, to prepare calibration and verification plans, and to protect the monitoring and measurement equipment.

Clause 7.1—Planning of product realization—prescribes that the organization must plan (according to the quality objectives and requirements for the product) and develop the processes needed for the product realization. The planning of the product realization must be consistent with the requirements of the other processes of the QMS. The organization shall also document one or more processes for risk management in product realization.

Customer-related processes are the focus of Clause 7.2 of this section. This concerns the capability of the manufacturer to determine the requirements related to the product, review them and communicate with customers the relevant aspects related to the production.

Clause 7.3—Design and development—focuses on tasks related to design and development. In particular, it requires that the organization plans and controls the design and development of the product. This clause also defines design and development inputs and outputs. The review, verification, and validation of product artifacts are also recorded. Finally, this clause presents requirements concerning design and development transfer, the control of design and development changes and design and development files.

Clause 7.4 focuses on purchasing processes and verification of purchased products.

Production and service provision falls within the scope of Clause 7.5, which includes 11 subclauses related to (1) the control of production and service provision, (2) the cleanliness of the product, (3) installation activities, (4) servicing activities, (5) particular requirements for sterile medical devices, (6) the validation of processes for production and service provision, (7) particular requirements for the validation of processes for sterilization, (8) identification (procedures for product identification throughout the product realization process), (9) traceability, (10) customer property, and (11) the preservation of the product.

The section concludes with Clause 7.6—Control of monitoring and measuring equipment—that indicates that the organization should prepare the monitoring and measurement to be undertaken and the monitoring and measuring equipment necessary to provide evidence of conformity of the product to the requirements.

5.2.5 Measurement, analysis, and improvement

Section 8 focuses on how to improve not only the product, but also the manufacturing process. Next, the section addresses how to improve the QMS.

Clause 8.1 introduces some general requirements concerning the planning and implementation of the monitoring, measurement, analysis, and improvement processes.

Monitoring and measurement (Clause 8.2) includes six subclauses that regulate: (1) a documented procedure for feedback, it being necessary to consider feedback as a potential input into the risk management of the product; (2) a documented procedure for timely complaint handling in accordance with any applicable regulatory requirements; (3) a documented procedure for providing notification to the appropriate regulatory authorities (if applicable) in the case of any adverse events; (4) the conduct of internal audits at planned intervals; (5) the application of suitable methods for the monitoring and, as appropriate, measurement of the QMS processes; and (6) the monitoring and measurement of the characteristics of the product.

Clause 8.3 focuses on the control of any nonconforming product. In particular, the organization must define actions in response to nonconforming products detected either before delivery or after delivery. Subsequently, the organization must perform a reworking.

The organization shall define documented procedures for data analysis (Clause 8.4) and improvement (Clause 8.5) to ensure and maintain the continued suitability, adequacy, and effectiveness of the QMS, as well as medical device safety and performance, through the use of the quality policy, quality objectives, audit results, postmarket surveillance, analysis of data, corrective actions, preventive actions, and management review. The CAPA model (Corrective Action, Preventive Action) is assumed.

5.3 ISO 13485:2016 versus other Quality Systems

Although ISO 13485 is a standard of the family ISO 9001 (Quality Management Systems), with a common intent and scope, it is a stand-alone standard that contains additional requirements which are specific for organizations involved in any stage of the Medical Device life cycle. Other elements of ISO 9001, not relevant as regulatory requirements, have been removed. ISO 13485, however, is designed to be integrated into other organization management systems.

It is important to highlight the differences between the latest version (2016) and the previous one (2003), since organizations currently compliant with the 2003 version are required to shift to the 2016 version by 2019.

The 2016 version of ISO 13485 places greater emphasis on risk management and risk-based decision making. The focus is 2-fold. On the one hand, the Standard concentrates more attention on risks associated with the safety and performance of Medical Devices. On the other hand, compliance with regulatory requirements is stressed throughout the Standard. In addition, ISO 3485:2016 requires that organizations apply more stringent controls when it comes to outsourcing processes.

The latest version of the Standard also reflects the increased regulatory requirements for all organizations acting along the Medical Devices supply chain.

Noteworthy points include:

- a greater emphasis on appropriate infrastructure such as for the production of sterile Medical Devices;
- an increased conformance with regulatory requirements and, particularly, with regulatory documentation;
- a greater attention on postmarket activities, including complaint handling and regulatory reporting;
- a broadening of the standard's range of application to encompass organizations that interact with the medical devices manufacturer, including activities such as (1) the supply of raw materials; (2) the design, development, correction, and maintenance of the Medical Device; (iii) the performance of services (e.g., sterilization); and (4) the import and distribution of Medical Devices;
- additional requirements in the design and development of the Medical Device, taking into account usability issues and human factors, the use of other standards affecting the Medical Device life cycle, and a more specific planning for the verification and validation of the design, prototype, and final product; and
- harmonization of validation requirements for different software applications, such as the QMS software, process control software, and the software for monitoring and measurement.

Another relevant QMS is the one defined by FDA 21 CFR Part 820.

In contrast to ISO 13485:2016, FDA 21 CFR Part 820 is the law for medical device manufacturers and vendors in the USA. In other words, a medical device company that would like to operate in the USA market is obliged to implement a QMS compliant with the FDA regulations.

Although the latest version of ISO 13485 has reduced the degree of disparity between the two standards, there are still some significant differences.

ISO 13485:2016 has reinforced the requirement for a risk-based approach to establishing and maintaining the QMS. FDA 21 CFR Part 820, instead, does not explicitly define risk-based requirements. However, the application of risk-based approaches is consistent with the assumptions of the FDA.

ISO 13485:2016 has introduced requirements for the protection of confidential health information and the addressing of the deterioration and loss of documents. For its part, FDA Part 820 has foundational requirements for documentation control and records management. These changes in the ISO 13485 requirements render the Standard more closely in accordance with the expectations of FDA 21 CFR 820.5 (Quality System), 820.40 (Document Controls), and 820.180 (Records).

In certain other cases, the changes introduced by the 2016 version of the standard have positioned it nearer to FDA CFR 820. Concerning human resources, for example, ISO 13485:2016 requires specific processes to establish competence and provide training, and in fact exceeds the provisions of the FDA in part 820.25. Other clauses that follow a similar trend are those related to design and development.

To conclude, the correspondence between the two sets of regulations is shown in Figure 5.3.

FDA 21 CFR Part 820 versus ISO 13485:2016

FDA QSR (21 CFR Part 820)	**ISO 13485:2016**
820.1 Scope	1 Scope
	2 Normative References
820.3 Definitions	3 Terms and Definitions
820.5 Quality System	4 Quality Management System
	4.1 General Requirements
	4.2 Documentation Requirements
820.20 Management Responsibility	5.0 Management Responsibility
820.20(a) Quality Policy	5.3 Quality Policy
820.20(b) Organization	4.1 Management Responsibility – General
820.20(b)(1) Responsibility & Authority	5.5 Responsibility & Authority
820.20(b)(2) Resources	5.1e Management Commitment
820.20(b)(3) Management Representative	5.5.2 Management Representative
820.20(c) Management Review	5.6 Management Review
820.20(d) Quality Planning	5.4 Quality Planning
820.20(e) Quality System Procedures	4.2.1 General
	4.2.2 Quality Manual
820.22 Quality Audit	8.2.4 Internal Quality Audits
820.25 Personnel	6 Resource Management
820.25(a) General	6.1 Provision of Resources
	6.2 Human Resources
820.25(b) Training	6.2 Human Resources
820.30 Design Controls	7.3 Design and Development
820.30(a) General	7.3 Design and Development
820.30(b) Design and Development Planning	7.1 Planning of Product Realization
	7.3.2 Design and Development Planning
820.30(c) Design Input	7.2.1 Customer-Related Processes
	7.2.2 Review of Requirements Related to Product
	7.3.3 Design and Development Inputs
820.30(d) Design Output	7.3.4 Design and Development Outputs
820.30(e) Design Review	7.3.5 Design and Development Review
820.30(f) Design Verification	7.3.6 Design and Development Verification
820.30(g) Design Validation	7.3.7 Design and Development Validation
820.30(h) Design Transfer	7.3.8 Design and Development Transfer
820.30(i) Design Changes	7.3.9 Control of Design and Development Changes
820.30(j) Design History File	7.3.10 Design and Development Files

Figure 5.3 (Continued)

FDA 21 CFR Part 820 versus ISO 13485:2016

FDA QSR (21 CFR Part 820)	ISO 13485:2016
820.40 Document Controls	4.2.4 Control of Documents
820.40(a) Document Approval and Distribution	4.2.4 Control of Documents
820.40(b) Document Changes	4.2.4 Control of Documents
820.50 Purchasing Controls	7.4.1 Purchasing Process
820.50(a) Evaluation of Suppliers, Contractors, and Consultants	7.4.1 Purchasing Process
820.50(b) Purchasing Data	7.4.2 Purchasing Information 7.4.3 Verification of Purchased Product
820.60 Identification	7.5.8 Identification
820.65 Traceability	7.5.9 Traceability
820.70(a) Production and Process Controls	7.5.1 Control of Production and Service Provision 7.5.6 Validation of Processes for Production and Service Provision
820.70(b) Production and Process Changes	6.3 Infrastructure 6.4 Work Environment and Contamination Control 7.5.1 Control of Production and Service Provision 7.5.6 Validation of Processes for Production and Service Provision
820.70(c) Environmental Control	6.4 Work Environment and Contamination Control
820.70(d) Personnel	6.2 Human Resources
820.70(e) Contamination Control	6.4.2 Contamination Control
820.70(f) Buildings	6.3 Infrastructure
820.70(g) Equipment	6.3 Infrastructure 7.5.1 Control of Production and Service Provision 7.5.6 Validation of Production and Service Provision
820.70(h) Manufacturing Material	7.5.11 Preservation of Product
820.70(i) Automated Processes	6.3.b Infrastructure 7.5.6 Validation of Production and Service Provision
820.72 Inspection, Measuring, and Test Equipment	7.6 Control of Monitoring and Measurement Equipment
820.75 Process Validation	7.5.6 Validation of Production and Service Provision

Figure 5.3 (Continued)

FDA 21 CFR Part 820 versus ISO 13485:2016

FDA QSR (21 CFR Part 820)	**ISO 13485:2016**
820.80(a) Receiving, In-process, and Finished Device Acceptance—General	7.1 Planning of Product Realization 7.4.3 Verification of Purchased Product 7.5.1 Control of Production and Service Provision
820.80(b) Receiving Acceptance	7.4.3 Verification of Purchased Product
820.80(c) In-Process Acceptance	7.1 Planning of Product Realization
820.80(d) Final Acceptance Activities	7.1 Planning of Product Realization
820.80(e) Final Acceptance Records	7.1 Planning of Product Realization
820.86 Acceptance Status	7.5.8 Identification
820.90(a) Nonconforming Product	8.3 Control of Nonconforming Product
820.90(b) Nonconformity Review and Disposition	8.3 Control of Nonconforming Product
820.100 Corrective and Preventative Action	8.5.2 Corrective Action 8.5.3 Preventative Action
820.120 Device Labeling	4.2.3 Medical Device File 7.5.8 Identification 7.5.11 Preservation of Product
820.130 Device Packaging	4.2.3 Medical Device File 7.5.8 Identification 7.5.11 Preservation of Product
820.140 Handling	4.2.3 Medical Device File 7.1 Planning of Product Realization 7.5.8 Identification 7.5.11 Preservation of Product
820.150 Storage	4.2.3 Medical Device File 7.1 Planning of Product Realization 7.5.8 Identification 7.5.11 Preservation of Product
820.160 Distribution	4.2.3 Medical Device File 7.1 Planning of Product Realization 7.5.8 Identification 7.5.11 Preservation of Product
820.170 Installation	4.2.3 Medical Device File 7.5.3 Installation Activities 7.5.8 Identification 7.5.11 Preservation of Product
820.180 Records	4.2 Documentation Requirements 4.2.3 Medical Device File 7.1 Planning of Product Realization
820.181 Device Master Record	4.2.3 Medical Device File

Figure 5.3 (Continued)

FDA 21 CFR Part 820 versus ISO 13485:2016

FDA QSR (21 CFR Part 820)	ISO 13485:2016
820.184 Device History Record	4.2.5 Control Records 7.1 Planning of Product Realization 7.5.8 Identification
820.186 Quality System Record	4.2 Documentation Requirements 7.1 Planning of Product Realization
820.198 Complaint Files	7.2.3 Communication 8.2.1 Feedback 8.2.2 Complaint Handling 8.2.3 Reporting to Regulatory Authorities
820.200 Servicing	4.2.3 Medical Device File 7.1 Planning of Product Realization 7.5.4 Servicing 7.5.8 Identification
820.250 Statistical Techniques	8.1 General 8.4 Analysis of Data

Figure 5.3 ISO 13485:2016 versus FDA 21 CFA Part 820

5.4 ISO 13485 certification

Certification is not mandatory, and organizations can obtain the benefits of the standard without being certified. However, third-party certification can be a way of demonstrating to stakeholders and regulatory authorities that the organization meets the requirements.

The following steps may be useful to prepare for ISO 13485 certification:

1. Planning the Quality System
2. Meeting requirements
3. Implementing design controls
4. Documenting, recording, and training
5. Managing processes
6. Auditing for the certification

Step 1: Planning the Quality System: The requirements for quality planning are part of the Standard (Clause 5.4). The leading idea must be that the preparation of a Quality Manual is not sufficient to satisfy this requirement. The organization is asked to document quality plans for implementing changes to the QMS. As part of the Quality Plan, a certification body should be selected. In order assist such selection, the official European page lists possible candidate bodies based upon the product category.

Step 2: Meeting Regulatory Requirements: In order to define the Quality Plan, the target market must be identified. For example, if the organization would like to sell its products in the USA and European markets, the Quality Plan must take into account the requirements presented in both regulations in addition to those asserted by ISO 13485 itself. Moreover, the Quality Plan must also address the differences that exist between the QMS requirements defined by the two sets of regulations.

Step 3: Implementing Design Controls: The organization must implement the design controls established by Clause 7.3, which are Design Planning, Design Inputs, Design Outputs, Design Reviews, Design Verification, Design Validation, and Design Changes.

Step 4: Documenting, recording, and training: The Quality Manual must define the process interactions for the Quality System. The document control procedure should be the first procedure defined. Indeed, this will represent the foundation for the entire Quality System. Successively, all the record control and design control procedures should be defined and approved. Next, the organization can start documenting the training relating to these procedures. Once the organization has a documented training process, it is ready to start writing the remaining 19 procedures required by ISO 13485, plus the other procedures required by national regulations.

Step 5: Managing processes: A well-established way of managing processes is via the CAPA process, internal audits, and management reviews.

Step 6: Auditing for the certification: The certification process is a two-stage audit. It is conducted to verify the compliance of the quality system to the requirements. It is based on evidence and, therefore, records are needed to show that the systems have been fully implemented.

5.5 Use of ISO 13485 in each jurisdiction

ISO 13485 is recognized and used differently in the different jurisdictions. The following analysis has been reported by the International Medical Device Regulators Forum [26]:

Australia—*Therapeutic Goods Administration (TGA)*

In Australia, ISO 13485:2003 is formally recognized under the Conformity Assessment Standards Order (Standard for Quality Management Systems and Quality Assurance Techniques), as a standard for the manufacture of all kinds of medical devices that require a QMS for conformity assessment.

QMSs that comply with the ISO 13485:2003 standard are treated as complying with the relevant parts of the conformity assessment procedures for QMSs set out in the Therapeutic Goods (Medical Devices) Regulations 2002.

Before the end of the transition period the TGA anticipates updating the relevant Conformity Assessment Standards order to reference the newer ISO 13485:2016.

Brazil—*National Health Surveillance Agency (ANVISA)*

The Brazilian Good Manufacturing Practice (RESOLUÇÃO DA DIRETORIA COLEGIADA—RDC Nº16) has its own QMS, which, however, recognizes certain clauses from ISO 13485:2003.

Canada—*Health Canada (HC)*

ISO 13485:2013 has been harmonized to meet the requirements of the Canadian regulations. The standard applies to Medical Devices falling within Classes II, III, and IV.

China—*China Food and Drug Administration (CFDA)*

ISO 13485 is Mandatory and it is accepted as a proof of Chinese QMS requirements. It applies to Classes II and III.

Europe—*European Commission (EC)*

ISO 13485:2012—a harmonized version of ISO 13458:2003—is required for the CE mark and facilitates compliance with Medical Device Directive [93/42/EEC], In-Vitro Diagnostic Directive [98/79/EC] and Active Implantable Medical Devices Directive [90/835/EEC]. ISO 13485 must be applied for Classes I (Sterile or Measuring Function), IIa, IIb, and III.

Japan—*Ministry of Health, Labour and Welfare (MHLW) Pharmaceuticals and Medical Devices Agency (PMDA)*

The regulation is based on ISO 13485, which must be applied for Classes II—Designated and controlled—and Classes III and IV—Specially controlled.

Russia—*Russian Ministry of Health Roszdravnadzor*

In Russia, ISO 13485:2003 is a recognized standard.

The United States of America—the *US Food and Drug Administration (US FDA)*

The regulations framework in USA has its own QMS defined by the 21 CFR Part 820, which must be applied for Classes I, II, and III.

Chapter 6

ISO 14971: medical devices—application of risk management to medical devices

6.1 Introduction

ISO 14971 is an international standard for risk management specifically devised for the development of medical devices. The standard is intended to help manufacturers to establish a full risk-management process that includes the identification of hazards, the assessment of the related risks, and the implementation of risk control measures.

The first version of the standard was proposed in 2000. The second edition was published in 2007. In the European Union, there is a harmonized version of the standard, the EN ISO 14971:2012, which is built upon the international version ISO 14971:2007, but which also includes information specific to the directives in the European Union. Therefore, in the European Union, the harmonized version must be followed, whereas in the rest of the world, the valid version (at the time of writing of this book) is the 2007 version. There is some confusion due to the fact that the European Union version is dated 2012 and may appear newer when, in reality, it is just a harmonization of the international (2007) version.

ISO 14971 deals with processes for managing risks. The fundamental principle is that Medical Devices can expose the patient, but also the operator, the caregiver, other equipment and the environment to hazards, which can cause loss of or damage to something they value.

The risk basically embraces two components: (1) the probability of occurrence of harm and (2) the severity of the harm. One of the most critical characteristics of a risk is that different stakeholders may place a different value on the severity of the harm and even on the probability of its occurrence. All stakeholders have to understand that the adoption of a medical device involves some degree of risk and "absolute safety" is never achievable.

The acceptability of a risk to a stakeholder depends on the stakeholder's perception of the risk. One stakeholder may perceive the risk differently from another depending on whether or not exposure to the hazard appears to be involuntary, avoidable, or due to a lack of quality. The decision to use a medical device in the context of hazards associated with the medical device itself implies an assessment of the risks associated with these hazards, a control of these risks, and a monitoring of the effectiveness of that control.

ISO 14971 applies to all kinds of medical devices and may be adopted to manage risks throughout all stages of the product life cycle. It applies to design, production, and all postproduction activities including postmarket surveillance. Since risks can be introduced at any stage of the product life cycle and since such risks can generally be controlled at a completely different stage, ISO 14971 must be extended throughout the entire product life cycle.

It is important to highlight that this standard does not apply to clinical decision-making. Indeed, whether or not a medical device should be used within a particular clinical procedure is a choice that must be the responsibility of clinicians.

ISO 14971 introduces a risk management process and is organized in the following parts:

Part 1—Establishing the scope of the standard;
Part 2—Defining terms;
Part 3—Establishing a risk management framework;
Part 4—Performing the risk analysis;
Part 5—Evaluating risk for each identified hazardous situation.
Part 6—Developing risk control measures when risks are unacceptable;
Part 7—Evaluating the residual risk posed by risk control measures;
Part 8—Reviewing risk management process and reporting on the review;
Part 9—Monitoring device during production and postproduction.

The process is detailed in Figure 6.1.

6.2 Contents

In ISO 14971:2007, several terms are either defined or reported from other standards. Among these, it is worth mentioning:

Harm: physical injury or damage to the health of people, or damage to property or the environment [*Source:* ISO/IEC Guide 51:1999, definition 3.3].

Hazard: potential source of harm [*Source:* ISO/IEC Guide 51:1999, definition 3.5].

Hazardous situation: circumstance in which people, property, or the environment are exposed to one or more hazard(s) [*Source:* ISO/IEC Guide 51:1999, definition 3.6].

Intended use, intended purpose: use for which a product, process, or service is intended according to the specifications, instructions, and information provided by the manufacturer.

Life cycle: all phases in the life of a medical device, from the initial conception to final decommissioning and disposal.

Objective evidence: data supporting the existence or verity of something [*Source:* ISO 9000:2005, definition 3.8.1].

Risk: combination of the probability of occurrence of harm and the severity of that harm [*Source:* ISO/IEC Guide 51:1999, definition 3.2].

Residual risk: risk remaining after risk control measures have been taken.

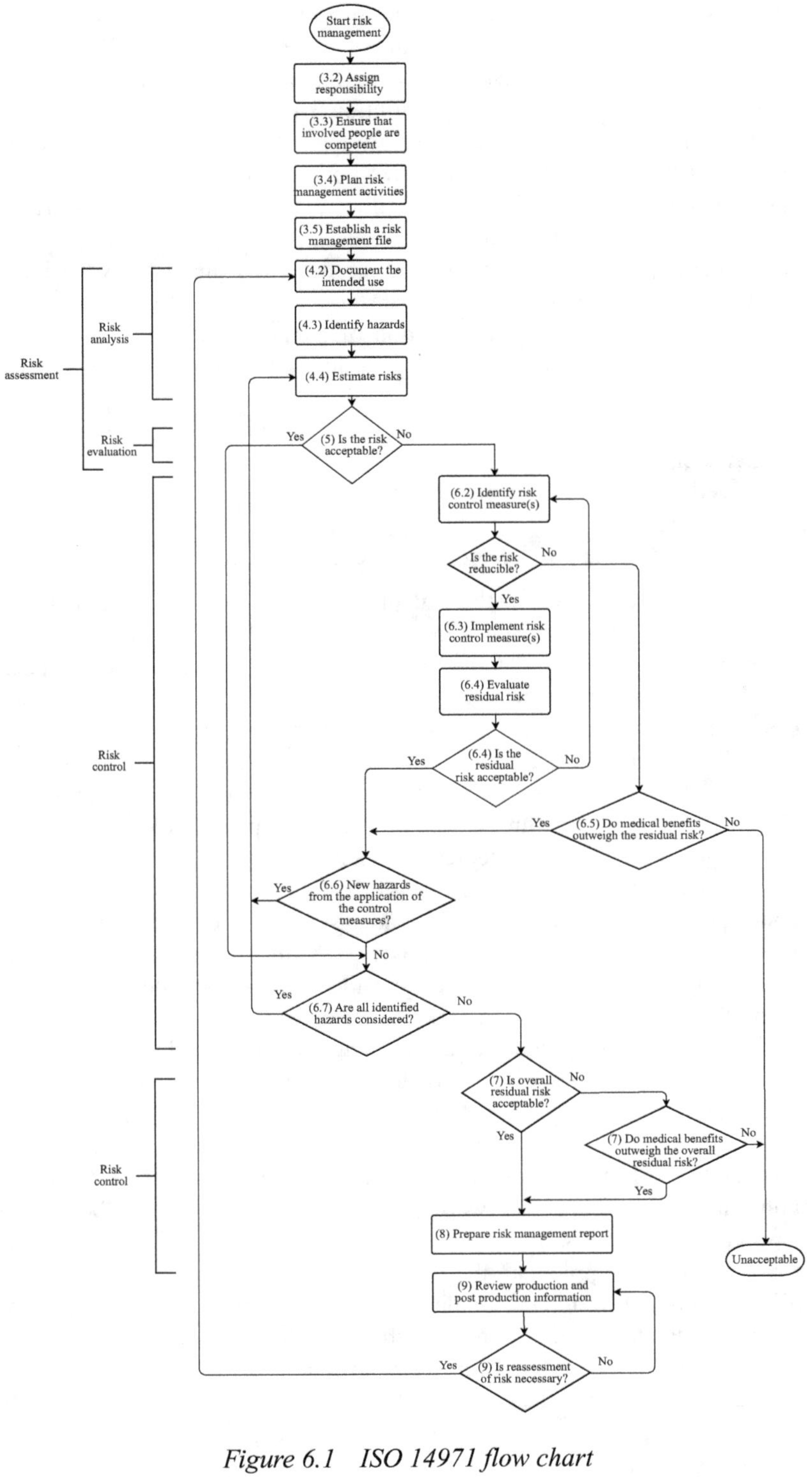

Figure 6.1 ISO 14971 flow chart

Risk analysis: systematic use of available information to identify hazards and to estimate the risk [*Source:* ISO/IEC Guide 51:1999, definition 3.10].

Risk assessment: overall process comprising a risk analysis and a risk evaluation [*Source:* ISO/IEC Guide 51:1999, definition 3.12].

Risk control: process in which decisions are made and measures implemented by which risks are reduced to, or maintained within, specified levels.

Risk estimation: process used to assign values to the probability of occurrence of harm and the severity of that harm.

Risk evaluation: process of comparing the estimated risk against given risk criteria to determine the acceptability of the risk.

Risk management: systematic application of management policies, procedures, and practices to the tasks of analyzing, evaluating, controlling, and monitoring risk.

Risk management file: set of records and other documents that are produced by risk management.

Safety: freedom from unacceptable risk [*Source:* ISO/IEC Guide 51:1999, definition 3.1].

Severity: measure of the possible consequences of a hazard.

Concerning the requirements, the structure of the standard is shown in Figure 6.2.

Part 3 of ISO 14971 introduces briefly the general risk management requirements, which are successively detailed in Parts 4–9. A process for the management and control of the risks associated with the organization's medical devices must be established. This part of the standard also asks the manufacturer to assign responsibility for risk management. Moreover, a policy that governs and controls how risk acceptability criteria are established must be defined. Finally, the general risk management requirements establishing a risk-management plan and keeping up to date the risk management file for the medical device.

Part 4 of the standard requires the use of the risk management file to document the intended purpose of the medical device and to specify the characteristics that could affect safety. This part of the standard imposes on the manufacturer the responsibility of carrying out a risk analysis based on the identification of the hazards and the estimation of the risk associated with the corresponding hazardous situations.

Part 5 requires the manufacturer to evaluate risks for each hazardous situation. The risk acceptability criteria, previously defined, must be used to decide if any risk reduction is necessary.

Part 6 regulates the identification of control measures to reduce unacceptable risks to an acceptable level of concern. Control measures must be implemented and their application must be verified. Afterwards, a new risk evaluation must be performed with respect to residual risks. In any case in which the residual risk is unacceptable and further risk control is impractical, the manufacturer must determine whether or not the medical benefits outweigh the residual risk (this requires a risk–benefit analysis). This part of the standard also requires that the manufacturer reviews the risk control measures because these could inadvertently introduce new hazards. All hazards and all hazardous situations must be considered.

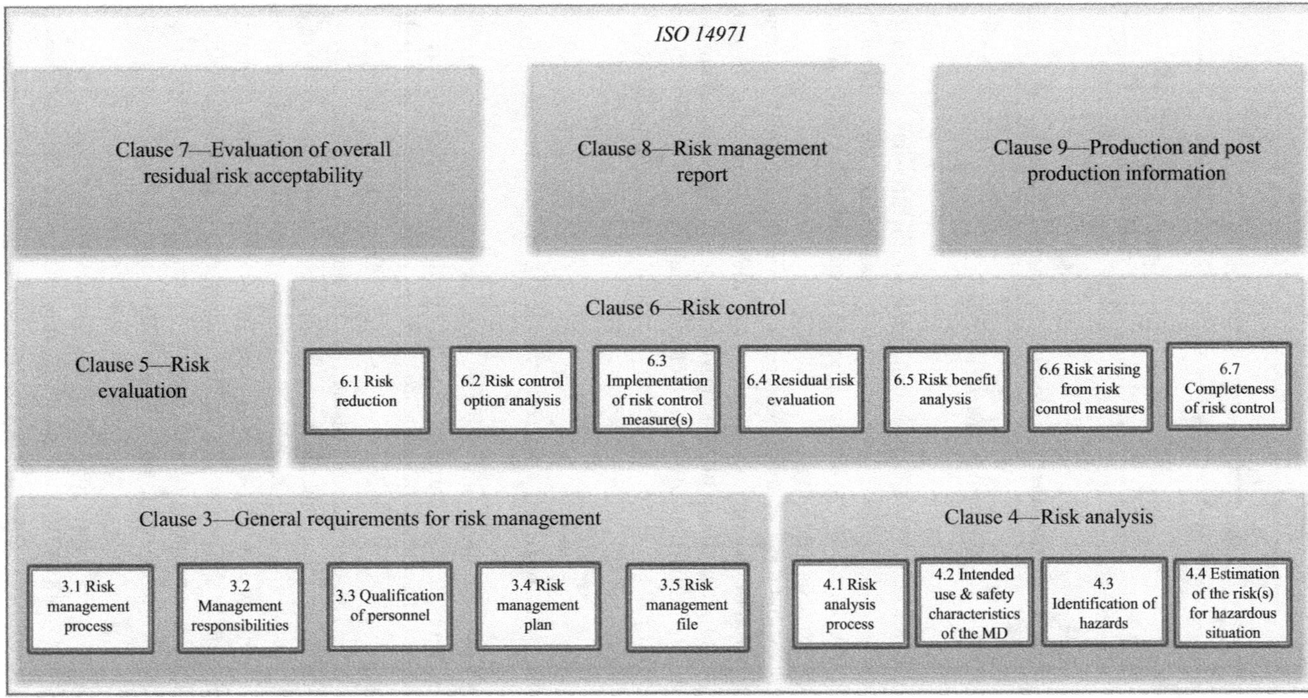

Figure 6.2 ISO 14971

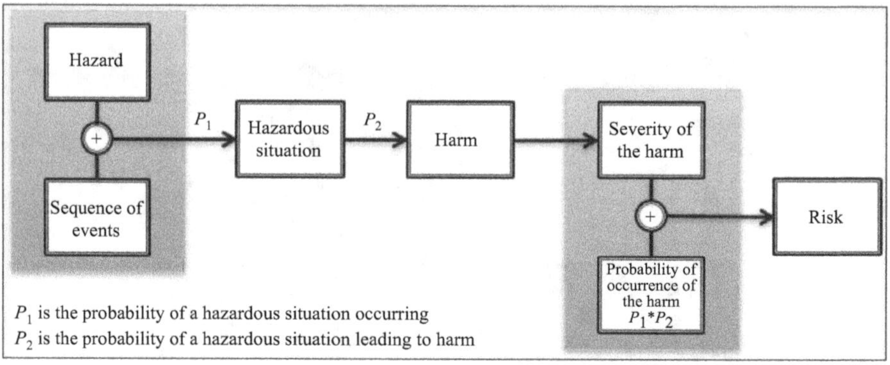

Figure 6.3 Risk components

Part 7 imposes the requirement by which the manufacturer has to evaluate the acceptability of the overall residual risks of the medical device. If any of such risks are unacceptable, it is mandatory to use evidence to decide whether or not the clinical benefits outweigh the residual risks.

Part 8 prescribes that a risk management review is performed and a risk management report is prepared before the medical device is put on the market. This part also requires that the risk management plan, and overall residual risks are reviewed.

Finally, Part 9 requires that the manufacturer establishes a system to monitor the performance of the medical device during production and postproduction phases (i.e. postmarket surveillance).

6.3 Risk concepts applied to medical devices

A risk is assessed by estimating both the severity and probability of occurrence of the harm that could result. The harm may be the result of a hazard when a certain sequence of events occurs. While the severity of the risk is estimated considering the potential harm, the probability of occurrence is related only to the occurrence of such a sequence of events. The components of the risk are depicted in Figure 6.3.

It is important to note that hazardous situations can arise even when there are no faults. However, for the aim of this book, we are interested in hazardous situations caused by software/hardware faults. It is also useful to clarify that the probability of a fault is not the same as the probability of the occurrence of the harm. A fault does not always result in a hazardous situation (it may remain latent or be masked by other internal conditions of the system), and a hazardous situation does not always result in harm.

There are two types of faults: random and systematic.

An example of a random fault is the failure of a part such as an integrated circuit in an electronic assembly.

A systematic fault can be caused by a human error in any activity of design, construction, and maintenance. It will systematically lead to failure when some specific combinations of input and internal conditions are in order. Otherwise, it will remain latent. Systematic faults may occur in both hardware and software layer and can be introduced at any time during the system life cycle. An example of a systematic fault is the type mismatch between an input value type expected by an interface and its corresponding field in the database when these two do not match each other. Any attempt at storing such information will lead to an exception, a truncation or other incongruences depending on the type of the mismatch.

There exist many methods and tools for the estimation of the risk. However, ISO 14971 does not prescribe any particular technique. Quantitative risk estimation is, of course, preferable. However, very often data are not available for performing this kind of evaluation. Therefore, a qualitative estimation is carried out.

Since a risk is the combination of the probability of occurrence of the harm and of its possible severity, risk estimation should examine factors such as the initiating circumstance, the sequence of events that could lead to the hazardous situation, the probability that such a situation arises and leads to the harm, and the nature of the harm.

In order to analyze a risk, its components (probability and severity) should be analyzed separately. Generally, when the harm is minimal or when the probability of occurrence is negligible, it will not be necessary to go beyond an initial hazard and consequence analysis.

In any case in which it is necessary to estimate and evaluate the risk, and data for a quantitative categorization are not available, the manufacturer should provide a qualitative description. To achieve this aim, risk acceptability criteria are previously defined. An example of probability and severity layers is reported in Table 6.1. In accordance with these layers, the risk matrix shown in Figure 6.4a, which is useful for later decision-making, is defined. The risk acceptability criteria are implemented in the matrix of Figure 6.4b where three ranges are represented: bottom-left corner for low risks, top-right corner for high (unacceptable) risks. The matrices of Figure 6.4c and 6.4d report, respectively, some risks as they are estimated before and after the application of control measures. In the case shown, the unacceptable risk (R_2) has been reduced to R_2 by the application of a particular control measure, but such a measure has also introduced a new risk (R_5), which is a possibility when applying a control measure.

Although probability is a continuous variable, in practice a discrete number of levels can be used, such as in the example reported in the previous figure. Several approaches are employed to estimate probabilities. Among these, the use of relevant historical data, the prediction of probabilities using analytical or simulation techniques, the use of experimental data, the reliability estimates, the production data, the postproduction information, and the use of expert judgment are commonly used. These approaches can be adopted individually or jointly. Wherever the use of multiple approaches is possible, they work as independent checks and this might help to increase confidence in the results. When these approaches cannot be used or are not sufficient, it might be necessary to rely only on expert judgment.

Typically, the probabilities of systematic faults, such as those leading to software failures, are extremely difficult to estimate. When the accuracy of the probability

Table 6.1 Examples of qualitative severity and probability levels

Qualitative severity levels	
Catastrophic	Results in patient death
Critical	Results in permanent impairment or life-threatening injury
Serious	Results in injury or impairment requiring professional medical intervention
Minor	Results in temporary injury or impairment not requiring professional medical intervention
Negligible	Results in inconvenience or temporary discomfort
Semiqualitative probability levels	
Frequent	$\geqslant 10^{-3}$
Probable	$< 10^{-3}$ and $\geqslant 10^{-4}$
Occasional	$< 10^{-4}$ and $\geqslant 10^{-5}$
Remote	$< 10^{-5}$ and $\geqslant 10^{-6}$
Improbable	$< 10^{-6}$

estimate is in doubt, it is often necessary to set a worst case with the highest probability.

In the absence of any data about the probability of occurrence of the harm, it is not possible to estimate accurately any risk and it is necessary to evaluate the risk from the nature of the harm alone. If it can be concluded that the hazard is of little consequence, the risk can be assessed as acceptable and no control measure is necessary.

It is usually assumed that there is an inverse relationship between the rigor of the processes used in the design and development of complex systems and the probability of systematic faults being introduced or remaining undetected. It is often appropriate to determine the required rigor of the development process and testing activities by taking into account the severity of the consequence of the systematic faults. The worse is the consequence and the less the effect of external control measures, the greater is the required rigor of the development process.

Concerning the severity of the potential harm, as for the probability, it is a continuous variable. However, in practice, the use of a discrete number of severity levels eases the analysis. In such cases, the manufacturer establishes how many categories these are and how they are to be defined. He/she uses descriptors appropriate to the levels in question, such as, for example, causes death, requires hospitalization, requires medical care, or does not require medical care. The severity levels will be chosen and justified by the manufacturer under clearly defined conditions of use.

6.4 Examples of hazards, foreseeable sequences of events and hazardous situations

The standard provides a nonexhaustive list of hazards that can be associated with different types of medical devices and a list of initiating circumstances that can result in

	Negligible	Minor	Serious	Critical	Catastrophic
Frequent					
Probable					
Occasional					
Remote					
Improbable					

(a)

	Negligible	Minor	Serious	Critical	Catastrophic
Frequent					
Probable					
Occasional					
Remote					
Improbable					

(b)

	Negligible	Minor	Serious	Critical	Catastrophic
Frequent		R_3			
Probable					
Occasional					R_2
Remote	R_1	R_4			
Improbable					

(c)

	Negligible	Minor	Serious	Critical	Catastrophic
Frequent		R_3			
Probable					
Occasional		R'_2		R'_2	
Remote	R'_1	R'_4		R'_5	
Improbable					

(d)

Figure 6.4 Impact–probability matrix: (a) matrix for severities and probabilities of Table 6.1, (b) example of risks estimation, (c) example of risk evaluation, (d) example of risk control

hazardous situations, which may be useful at the stage of the risk analysis concerning the identification of hazards, sequences of events and harms. Table 6.2 reports some of the possible hazards described in the standard.

In order to identify foreseeable sequences of events, it is often useful to consider the initiating circumstances that can cause them. Table 6.3 provides examples of initiating events and circumstances, organized into categories. The list is not exhaustive, but it is intended by the Technical Committee to demonstrate several different types

Table 6.2 Some examples of hazards

Examples of operational hazards	
Function	Incorrect or inappropriate output or functionality
	Incorrect measurement
	Erroneous data transfer
	Loss or deterioration of function
User error	Attentional failure
	Memory failure
	Rule-based failure
	Knowledge-based failure
	Routine violation
Examples of information hazards	
Labeling	Incomplete instructions for use
	Inadequate description of performance characteristics
	Inadequate specification of intended use
	Inadequate disclosure of limitations
Operating instructions	Inadequate specification of accessories to be used with the MD
	Inadequate specification of preuse checks
	Over-complicated operating instructions
Warnings	Warnings of side effects
	Warnings of hazards likely with reuse of single-use MDs

of initiating events and circumstances that need to be taken into account to identify the foreseeable sequences of events for a medical device.

6.5 Risk-management methods and tools

In the following section, a general overview of some of the most commonly used tools that might be adopted in risk management by industry is reported. This is not an exhaustive list of tools that a manufacturer can use to manage risks. A more comprehensive list is available in [27]. Nor are such tools mutually exclusive. Indeed, depending on the situation, a combination of tools may be adopted.

6.5.1 Failure mode effects analysis

Failure mode effects analysis (FMEA) [28] enables the evaluation of potential failure modes for processes and components and their possible effects.

FMEA is a systematic and proactive approach for the identification of the potential ways in which a process or a component can fail, why it might fail, and how it can be made safer.

To realize an FMEA, the analyst identifies the components of the system, which are subsequently divided into subcomponents. For each subcomponent, failure modes

Table 6.3 Some examples of initiating events and circumstances

Category	Examples of initiating events and circumstances
Incomplete requirements	Inadequate specification of: – design parameters – operating parameters – performance requirements – in-service requirements (e.g. maintenance, reprocessing) – end of life
Manufacturing processes	Insufficient control of changes to manufacturing processes Insufficient control of materials compatibility information Insufficient control of manufacturing processes Insufficient control of subcontractors
Human factors	Potential for use errors triggered by design flaws, such as – confusing or missing instructions for use – complex or confusing control system – ambiguous or unclear device state – ambiguous or unclear presentation of settings, measurements, or other information – misrepresentation of results – insufficient visibility, audibility, or tactility – poor mapping of controls to actions, or of displayed information to actual state – controversial modes or mapping as compared to existing equipment – use by unskilled/untrained personnel – insufficient warning of side effects – inadequate warning of hazards associated with reuse of single-use MDs – incorrect measurement and other metrological aspects – incompatibility with consumables/accessories/other medical devices – slips, laps, and mistakes
Failure modes	Unexpected loss of electrical/mechanical integrity Deterioration in function as a result of maintenance for software Fatigue failure

are identified and their related effects are determined. The results are reported in a table where the rows describe the components/subcomponents and the columns represent respectively the failure mode, the possible cause, and the possible effect.

An extract of an FMEA for a tele-monitoring service, which is based on a Body Sensor Network, is reported in Table 6.4.

6.5.2 Failure mode, effects, and criticality analysis

Failure mode, effects, and criticality analysis (FMECA) [28] represents an extension of FMEA. Indeed, it includes an investigation of the degree of severity of the consequences, their respective probabilities of occurrence, and their detectability.

Table 6.4 An example of a Failure Mode Effects Analysis

Component	Subcomponent	Potential failure mode	Potential effects of failure	Potential causes of failure
Sensor	Sensor board	Stuck at zero	The device is out-of-order; it does not deliver any output to inputs	Sensing hardware
		Null reading	The device delivers null output values	Sensing hardware
		Out of scale reading	The device delivers no meaningful values	Sensing hardware
	Power supply	Stuck at zero	The device is out-of-order; it does not deliver any output to inputs	Natural energy exhaustion
		Reset	The node resets itself to its initial conditions	Anomalous current request that cannot be supplied by batteries
Intra BAN communication	Transport and routing	Packet loss	The radio packet is not delivered	Packet corruption Buffer overrun
		Isolation	The node is no longer connected to the sink node	Failure of all forwarding nodes
	Bluetooth stack	Bluetooth stack failure	A Bluetooth module (e.g., L2CAP, BNEP, etc.) fails	Bluetooth stack corrupted state
	Bluetooth channel	Header corruption	Header delivered with errors	Packet corruption
		Header length mismatch	Header length deviates from the specified one	Packet corruption
		Payload corruption	Payload delivered with errors	Packet corruption
Extra BAN communication		Data delivery failures	The network is not able to deliver the required number of measurements	The number of failed nodes is greater than a given threshold
		Cellular/Wi-Fi network unavailable	Monitoring stopped	Area without cellular/Wi-Fi signal

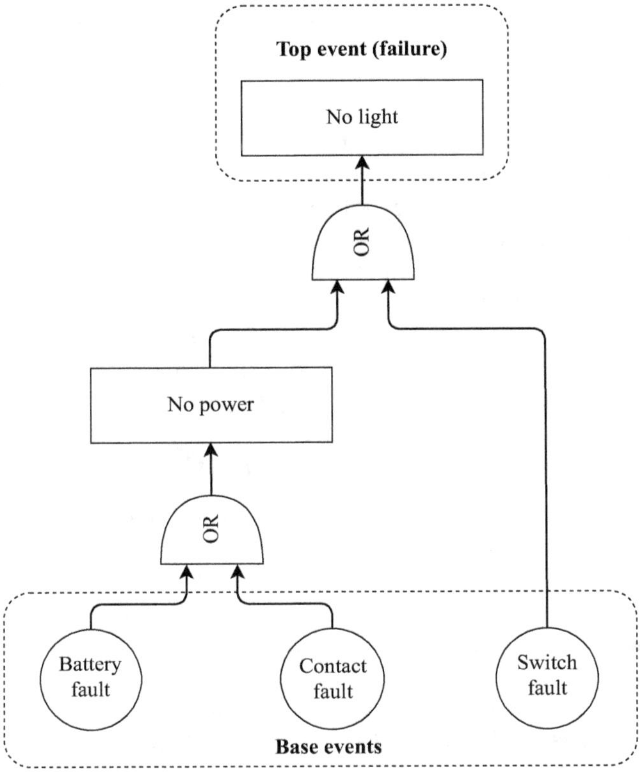

Figure 6.5 An example of a fault tree analysis

The result of an FMECA is a risk score, which may be used to rank the failure modes on a relative risk basis.

6.5.3 Fault tree analysis

Fault tree analysis (FTA) [29] is a systematic method of system analysis that examines a system from top down. This approach evaluates system (or subsystem) failures one at a time, but it enables the risk analyst to combine multiple causes of failure by identifying causal chains. Combinations of fault modes are described with logical operators and represented by means of trees. An example of such a representation is reported in Figure 6.5, where a simplified FTA analysis for a vehicle headlamp failure is reported.

6.5.4 Hazard analysis and critical control points

Hazard analysis and critical control points (HACCP) is a systematic, proactive, and preventive tool for assuring product quality, reliability, and safety.

HACCP is a process that comprises the following steps: (1) conducting a hazard analysis and identifying preventive measures for each step of the process; (2) determining the critical control points; (3) establishing critical limits; (4) establishing a system to monitor the critical control points; (5) establishing the corrective action to be taken when monitoring indicates that the critical control points are not in a state of control; (6) establishing a system to verify that the HACCP system is working effectively; and (7) establishing a record-keeping system.

As reported by the Food and Drug Administration (FDA), HACCP is traditionally used as "a management system in which food safety is addressed through the analysis and control of biological, chemical, and physical hazards from raw material production, procurement and handling, to manufacturing, distribution and consumption of the finished product." However, new fields of application of this tool are emerging, such as, for example, the pharmaceutical industry [30].

6.5.5 Hazard operability (HAZOP) analysis

Hazard operability (HAZOP) analysis [31] relies on the theory that assumes that risk events are due to deviations from the design or operating intentions. The analysis adopts a brainstorming technique for the identification of hazards using "guide-words" such as, for example, "No", "Other Than", "Part of", or "More", applied to relevant parameters to help identify potential deviations from normal use or design intentions.

HAZOP can potentially be applied to manufacturing processes, as well as in the pharmaceutical industry. The output of a HAZOP analysis is a list of critical operations for risk management.

6.5.6 Preliminary hazard analysis

Preliminary hazard analysis (PHA) [27] is a technique that relies on the use of prior knowledge and experience to identify potential hazards, hazardous situations and events that might cause harm. The approach consists of the following steps: (1) identifying the possibility that the risk event occurs, (2) evaluating qualitatively the possible injury or damage to health that could happen, (3) ranking the hazards on the basis of severity and likelihood of occurrence, and (4) identifying possible counter-measures.

PHA is quite a simple method that might be useful when circumstances prevent the use of a more extensive technique. For example, PHA can be used early in the development process when there is little information on the design details or system architecture. Therefore, hazards identified as a result of the PHA are further assessed with other risk assessment techniques.

6.5.7 Markov analysis

Markov analysis [27] is used to evaluate finite-state systems, where the future state depends only upon the present one. The approach can be extended to more complex systems by using higher order Markov processes. It is restricted only by the model, the mathematical computations and the assumptions.

		Current state		
		S1	S2	S3
Next state	S1	0.95	0.3	0.2
	S2	0.04	0.65	0.6
	S3	0.01	0.05	0.2

(a)

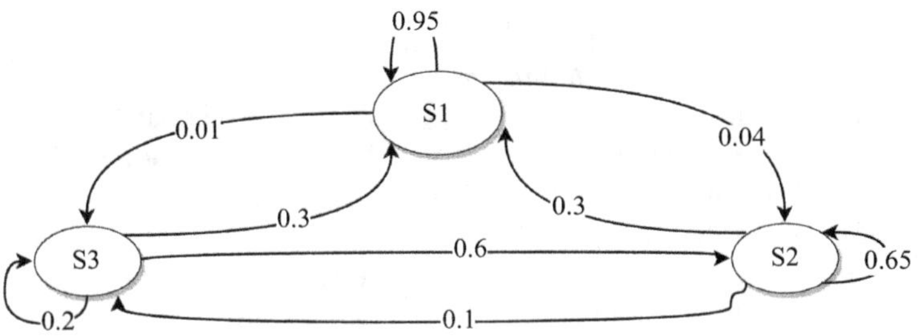

(b)

Figure 6.6 An example of a Markov analysis. (a) Example of a Markov matrix. (b) Example of a Markov diagram

It is a quantitative technique that uses probabilities of change between the states of the system. The Markov analysis can be automated by computer programs, such as model-checkers.

As an example of a Markov analysis, we report a system (analyzed in [27]) that can be in only three states: functioning (S1), degraded (S2), or failed (S3). The table in Figure 6.6a reports the transition probability. Figure 6.6b shows the Markov matrix and the equivalent diagram.

The Markov analysis can enable the computation of the probability (Pi) that the system is in the state Si. To do this, the following system of equations may be considered:

$$P_1 = 0.95P_1 + 0.30P_2 + 0.20P_3 \tag{6.1}$$

$$P_2 = 0.04P_1 + 0.65P_2 + 0.60P_3 \tag{6.2}$$

$$P_3 = 0.01P_1 + 0.05P_2 + 0.20P_3 \tag{6.3}$$

However, such equations are not independent. Consequently, one must be discarded and the following equation must be added:

$$P_1 + P_2 + P_3 = 1 \tag{6.4}$$

The solution is $P_1 = 0.85$, $P_2 = 0.13$, and $P_3 = 0.02$. The system is fully functioning for 85% of the time, in the degraded state for 13% of the time and failed for 2% of the time.

6.6 Use of ISO 14971:2007 in each jurisdiction

ISO 14971 is recognized and used differently in the different jurisdictions. The following analysis has been reported by the International Medical Device Regulators Forum [26]:

Australia—*Therapeutic Goods Administration (TGA)*
All medical devices are required to meet the Australian Essential Principles (EPs). The TGA's nonmandatory Medical Devices Standards Order (standards for risk management), 2008 (MDSO) specifies EN ISO/ISO 14971:2000 Clauses 1 to 9 inclusive or EN ISO/ISO 14971:2007 Clauses 1 to 9 inclusive to be used as a method to identify the risk associated with the use of the device (but not to be used as a specific means to implement the reduction of risks). Compliance with these standards is used as evidence of compliance with the EPs.

Brazil—*National Health Surveillance Agency (ANVISA)*
It is mandatory for manufacturers and importers of medical devices to have registries of risk management. ISO 14971:2007 may be employed in risk management reports. Several ANVISA regulations mention Standard 14971 and it is applicable to both pre- and postmarket stages.

Canada—*Health Canada (HC)*
In Canada, conformance to specific standards is not mandatory. However, evidence of conformance to recognized standards can be submitted to demonstrate that the specific requirements of the Medical Devices Regulations have been met. HC publishes a list of recognized standards, and the level of evidence expected is "equivalent or better" to these recognized standards. ISO 14971:2007 is currently a recognized standard, and represents an accepted approach to Risk Management. A risk analysis report is required for all Class IV medical device license applications, and for Classes III and IV Investigational Testing Authorization applications.

China—*China Food and Drug Administration (CFDA)*
The ISO14971:2007 has been translated into the China industry standard: YY/T 0316-2008 entirely and has been implemented since 2009. It is not mandatory standard but only as recommended standard. It is a very important standard for industry for risk management and for the preparation of the relevant documents for registration.

Europe—*European Commission (EC)*
The corresponding European standard EN ISO 14971:2012 is a European harmonized standard, which provides for a process to address the general risk management aspects

related to medical devices which are included in the legal requirements. However, it is not the primary goal of the standard to provide a direct presumption of conformance with any of the specific relevant European legal requirements on devices. Manufacturers and conformity assessment bodies need to feed the specific legal requirements on devices into the risk management process provided by the standard.

Japan—*Ministry of Health, Labour and Welfare (MHLW) Pharmaceuticals and Medical Devices Agency (PMDA)*
All medical devices are required to satisfy the EPs that align with those defined in GHTF/SG1/N68:2012 Essential Principles of Safety and Performance of Medical Devices. ISO 14971:2007 can be used for this purpose, which is clearly referred to in the checklist of EPs or the certification/approval standards for each medical device.

Russia—*Russian Ministry of Health Roszdravnadzor*
In the current regulation the use of standards is voluntary in the pre-market MD evaluation. Moreover, the Regulator does not recognize any standard which could provide a presumption of conformance. However, on the market, some types of MD have to be certified for particular mandatory standards (a list of mandatory standards and types of MD is available on the Regulator's web site). It should be noted, that this regulation was canceled on 01/01/2016.

The United States of America—*US Food and Drug Administration (US FDA)*
ISO 14971:2007 is recognized by the US FDA medical device program as a consensus standard in respect of which a person may submit a declaration of conformance in order to meet a premarket submission requirement or other requirements to which a standard is applicable. The US FDA by recognizing ISO 14971:2007 is acknowledging that the standard provides a framework to establish a risk-management system and processes that are an integral part of a manufacturer's quality management system and are applicable to all stages of the life cycle of the medical device. The principles of risk management discussed in the standard are considered suitable to assist in the management of the risks associated with the use of medical devices.

Chapter 7

IEC 62304: medical device software—software life-cycle processes

7.1 Introduction

IEC 62304 is a framework of life-cycle processes, with activities and tasks, which focuses on the design and maintenance of safe medical device software.

The fundamental principle is that several processes should be instantiated and implemented in the development and maintenance of software for medical purposes and that the choice of processes depends on the risks to the patient, clinical staff, caregivers and, more generally, other users. This is due to the consideration that relying only on software testing is usually not sufficient to build confidence in the product safety. The processes indicated in this standard may fall within two categories:

- processes concerning the risks arising from the operation of each software item of the system, which must be executed for all medical device software and
- processes aimed at achieving an appropriately low probability of software failure for each software item, chosen on the basis of the determined risks, which must be executed for a restricted selection of software items.

It is assumed that software for medical purposes is developed and maintained under the umbrella of a quality management system and a risk management system. ISO 14971 is a recognized standard by IEC 62304. Therefore, only some minor additional risk management requirements are presented by this standard.

The activities of the software development process are described in Clause 5. The software maintenance process is considered to be as important as the software development process because many incidents are related to service or maintenance including inappropriate software updates and upgrades. Consequently, the Standard poses stringent requirements, stressing the importance of maintenance activities, which are described in Clause 6. The Standard identifies two additional processes that are considered essential for the development of safe medical device software: the software configuration management process, which is the focus of Clause 8, and the software problem resolution process, which is presented in Clause 9.

The processes are summarized in Figure 7.1.

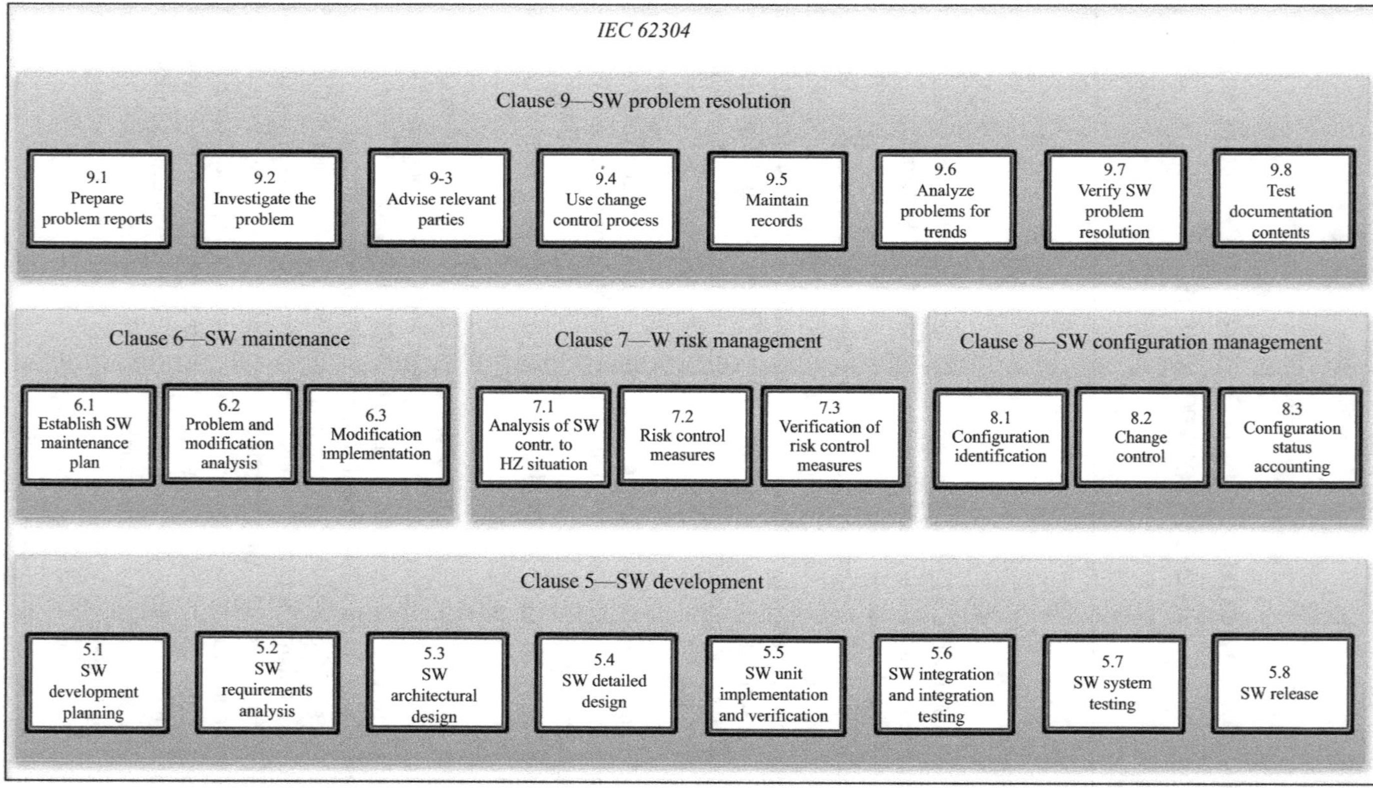

Figure 7.1 IEC 62304

7.2 Content

Hereafter, we refer to the 2006 version of the International Standard. IEC 62304:2006, after having introduced the Scope, Normative References, and Terms and Definitions, proposes with Clause 4 some General Requirements. In particular, Clause 4.1 requires that the manufacturer adopts a quality management system that consistently meets customer requirements and applicable regulatory requirements. Clause 4.2 (Risk Management) states that the manufacturer must apply a risk management process compliant with ISO 14971. The software safety classification is addressed in Clause 4.3. Indeed, three software safety classes (A, B, or C) are introduced according to the possible effects on the patient or other user resulting from a hazard connected to the software under consideration. Any software item is categorized accordingly to the following severity levels: Class A—No injury or damage to health is possible; Class B—Nonserious injury is possible; and Class C—Death or serious injury is possible.

The classification has to be documented and the level of the safety class assessment can be reduced through the application of risk control measures.

All clauses (100%) affect Class C software systems; whereas approximately 93% and 43% influence the development and maintenance of software systems falling into Class B and Class A, respectively.

7.2.1 Software Development Process

Clause 5 addresses the Software Development Process. In particular, Clause 5.1 requires the definition and update of a software development plan. The software development plan must include or reference procedures for coordinating the software design, development, and validation. The software development plan must also include, or reference or be linked to, other planning documents (i.e., verification planning, risk management planning, documentation planning, and software configuration management planning). In addition to such requirements, which are mandatory for software falling into any class of risk (A, B, or C), there are some additional requirements for those of type B or C (e.g., Software Integration Planning) and others concerning exclusively Class C software.

The software requirements analysis is regulated by Clause 5.2. This clause requires, for software of any class of risk, that the manufacturer defines and documents every requirement. A software requirement must include: (1) functional and capability requirements; (2) software system inputs and outputs; (3) interfaces between the software system and other systems; (4) software-driven alarms, warnings, and operator messages; (5) security requirements; (6) usability engineering requirements that are sensitive to human errors and training; (7) data definition and database requirements; (8) installation and acceptance requirements of the delivered medical device software at the operation and maintenance site or sites; (9) requirements related to methods of operation and maintenance; (10) user documentation to be developed; (11) user maintenance requirements; and (12) regulatory requirements. The software requirements must include risk control measures.

The next Clause (5.3) focuses on the Software Architectural Design. This concerns software items and systems falling into Class B or C and compels the manufacturer to transform the software requirements into a documented architecture, including the internal structure and the interfaces between the software items controlling the system, and to verify the resulting architecture. Specific attention is focused on SOUP (Software of Unknown Provenance) components, for which it is necessary to specify functional and performance requirements, as well as hardware and software platform needs. Only in the case of Class C systems does the manufacturer have to identify the segregation between software items.

Clause 5.4 focuses on the detailed software design. It requires that the software architecture is refined into software units and, only for Class C items, that a detailed design for both software units and interfaces is realized and verified.

Software unit implementation and verification is the subject of Clause 5.5. In particular, in addition to the software unit implementation, the clause compels the manufacturer, for Classes B and C, to establish both a software unit verification process and acceptance criteria (Class B and C). The software unit must be verified.

The next clauses (5.6–5.8) are mandatory for software items falling into Class B or C. Clause 5.6 regulates the software integration and integration testing. Clause 5.7 regulates the system level testing. This clause refers to the Software Problem Resolution process (of the standard itself) in any case in which an anomaly is detected. Finally, Clause 5.8 concerns the software release and emphasizes the need for a complete verification, evaluation, and documentation of known residual anomalies. The clause also requires (even for Class A items) the documentation of the released versions.

A naturally compliant software development process is the V-Model, that is described in detail in Chapter 12. Figure 7.2 shows the V-Model extrapolated from the Standard IEC 60601-1, which regulates the development of the Programmable Electrical Medical System (PEMS). Within the scope of IEC 62304, the "requirements for software are a subset of the requirements for a PEMS."

7.2.2 Maintenance process

The software maintenance process is regulated by Clause 6. Almost all the requirements apply to all classes of software (A, B, and C).

The manufacturer has to establish a preliminary software maintenance plan in order to manage the maintenance process, in accordance with Clause 6.1.

Clause 6.2 relates to the analysis of any problems and their rectification. It requires the documentation and evaluation of feedback, the use of the Software Problem Resolution process, the analysis and approval of any change request, and the communication to users and regulators.

Next, Clause 6.3 focuses on the implementation of the modification.

7.2.3 Software risk management process

Clause 7 compels the manufacturer to manage software risks. Almost all the requirements presented in this clause apply to software items of Class B or C. In particular,

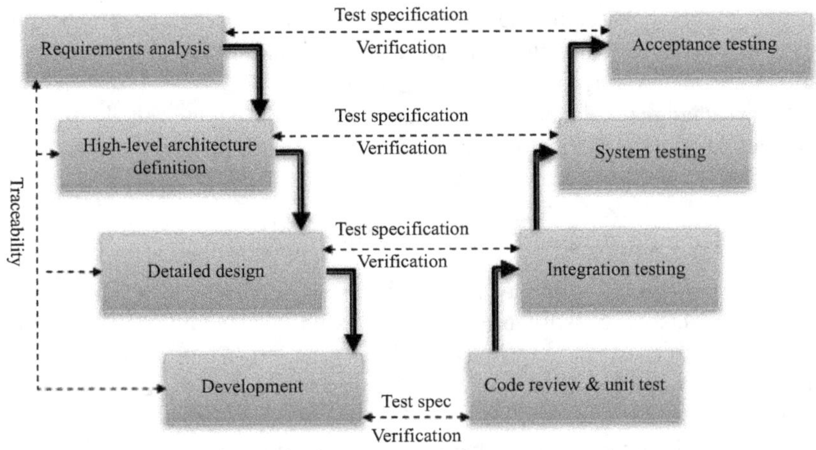

Figure 7.2 A V-Model development process compliant with IEC 62304

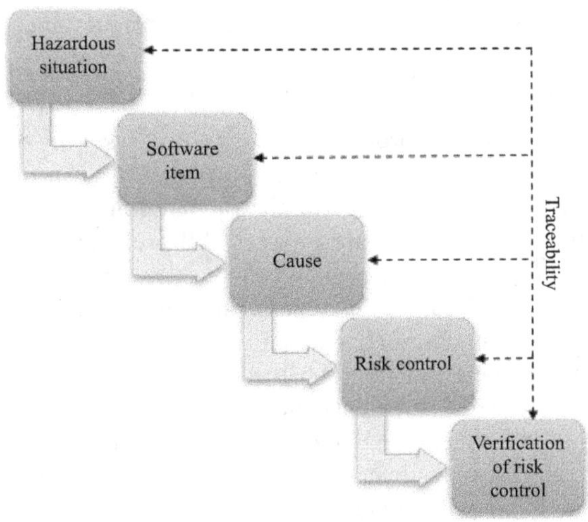

Figure 7.3 Traceability of software hazards

Clause 7.1 requires that the manufacturer analyzes the software contribution to hazardous situations. This means that software items that could contribute to hazardous situations identified within the Medical Device Risk Analysis Activity of ISO 14971 must be identified. The potential causes of the contribution must be identified and documented in a Risk Management File. Additionally, any sequences of events that could lead to hazardous situations must be documented.

Risk control measures are regulated by Clause 7.2. This concerns the definition of risk control measures and their implementation in software.

Clause 7.3 presents requirements relating to the verification of risk control measures. It includes the documented traceability of software hazards. In particular, the flow of dependency to trace is shown in Figure 7.3.

Software changes are addressed in Clause 7.4. In particular, the clause requires that changes to medical device software are analyzed with respect to safety and that risk management activities are performed.

7.2.4 Software configuration management process

The software configuration management process is regulated by Clause 8, which applies to all classes of software items (Classes A, B, and C). This process must include an activity for configuration identification (Clause 8.1) that requires that the manufacturer identifies the SOUP components and system configuration documentation.

Clause 8.2 focuses on change control. After any change request has been approved, this must be implemented and verified.

The last clause of the process (Clause 8.3) concerns configuration status accounting, which requires the manufacturer to retain retrievable records of the history of controlled configuration items including the system configuration.

7.2.5 Software problem resolution process

Clause 9 concerns the software problem resolution process. This process is mandatory for all classes of software items (Classes A, B, and C). The first activity of the process—defined by Clause 9.1—regards the preparation of a problem report for each problem detected in the software product. All problem reports must include information about the (1) type, which may be corrective, preventive, or adaptive to new environments; (2) scope, which includes characteristics like resources involved, time to change, and magnitude of change; and (3) criticality, that may be related to quality attributes like security, safety, robustness, etc.

The next step (Clause 9.2) is related to the investigation of the problem. The cause of the problem should be identified and any potential effects should be evaluated by implementing the risk management process. As a final result of the activity, a documented change request should be issued.

Subsequently, as defined by Clause 9.3, all interested parties must be notified and the change control process must be executed to approve and implement the change request (Clause 9.4). All problem reports must be recorded in accordance with Clause 9.5. Other requirements presented by Clause 9 concern the need for (1) an analysis to determine problems trends (Clause 9.6); (2) a verification of software problem resolution (Clause 9.7); and (3) test documentation (Clause 9.8).

7.3 Use of IEC 62304 in each jurisdiction

A study of the International Medical Device Regulators Forum has reported on the application of the IEC 62304 Standard in different jurisdictions [32]:

Australia—*Therapeutic Goods Administration (TGC)*
All medical devices are required to comply with the Australian Essential Principles (EPs). IEC 62304 (Software Life-Cycle Processes) and IEC 62366 (Usability Engineering) are referenced in the supporting data form and compliance with these standards is used as evidence of compliance with the EPs.

Brazil—*National Health Surveillance Agency (ANVISA)*
IES 62304:2006 may be employed either in a technical dossier or in technical reports. However, currently, it is not mandatory to be compliant with either of these two standards.

Canada—*Health Canada (HC)*
In Canada, conformance with medical standards is not mandatory. However, evidence of conformance with recognized standards can be submitted to demonstrate that the specific requirements of the Medical Devices Regulations have been met. HC publishes a list of recognized standards, and the level of evidence expected is "equivalent or better" to these recognized standards. IEC 62304:2006 is currently a recognized standard, and represents an accepted approach to the software development process for medical devices.

China—*China Food and Drug Administration (CFDA)*
IEC 62304:2006 had been translated into the China industry standard: YY/T 0664-2008 entirely and implemented since 2009. It is not a mandatory standard, but conformance is recommended.

Europe—*European Commission (EC)*
IEC 62304:2006 has been harmonized with the European standard EN 62304:2006. This standard provides a presumption of conformance with the legal requirements relating to the development life cycle for software as a medical device. The use of this standard provides one solution to ensure compliance with the legal requirements, although this can, however, be ensured also by other means.

Japan—*Ministry of Health, Labour, and Welfare (MHLW)*
IEC 62304:2006 has not been referred to so far, but it may be used for a premarket application process to satisfy the EPs that align with those defined in GHTF/SG1/N68:2012 Essential Principles of Safety and Performance of Medical Devices.
The United States of America—*US Food and Drug Administration (US FDA)*
IEC 62304:2006 is recognized as a consensus standard. It allows the submission of a declaration of conformance to meet a premarket submission requirement.

Chapter 8

IEEE 1012 and ISO/IEC 29119: standards for software verification

8.1 IEEE Std 1012 for system and software verification and validation

IEEE standard 1012 [33] focuses on verification & validation (V&V) processes and applies to systems, software, and hardware being developed, maintained, or reused (e.g., Commercial Off-the-Shelf (COTS) components). V&V processes include the analysis, evaluation, review, inspection, assessment, and testing of products. V&V may be performed at the level of the system, software element, or hardware element, or on any combination of these. V&V may also be performed on an element of a system, including a subordinate system (i.e., subsystem). In each case, the V&V processes are invoked, either in parallel or recursively, across the full life cycle of the system or element. The following key concepts are emphasized in this standard:

- the integrity levels, which describe the relevance of the system, software, or hardware, varying from high integrity to low integrity;
- the minimum number of V&V tasks for each integrity level, which specifies the minimum number of V&V tasks required for each integrity level. Some optional V&V tasks are also defined to address the project;
- the intensity and rigor applied to the V&V tasks, which assert that the intensity and rigor vary according to the integrity level. Any increase in integrity requires greater intensity and rigor in the V&V task;
- detailed criteria for the V&V tasks, which defines specific criteria for each V&V task (e.g., minimum criteria for correctness, consistency, completeness, accuracy, readability, and testability);
- systems viewpoints, which asserts the minimum software and hardware V&V tasks necessary to address system issues. These tasks include hazard and risk analysis, migration and retirement assessment, etc.;
- conformance to international and IEEE standards, which concerns the need to conform to life cycle process standards (e.g., ISO/IEC 15288:2008 and ISO/IEC 12207:2008) and to other IEEE software engineering standards.

V&V on software is addressed in Clause 7, which defines activities of V&V common to any level (system, software, or hardware), and Clause 9, which defines the activities specific to the software level (Figure 8.1).

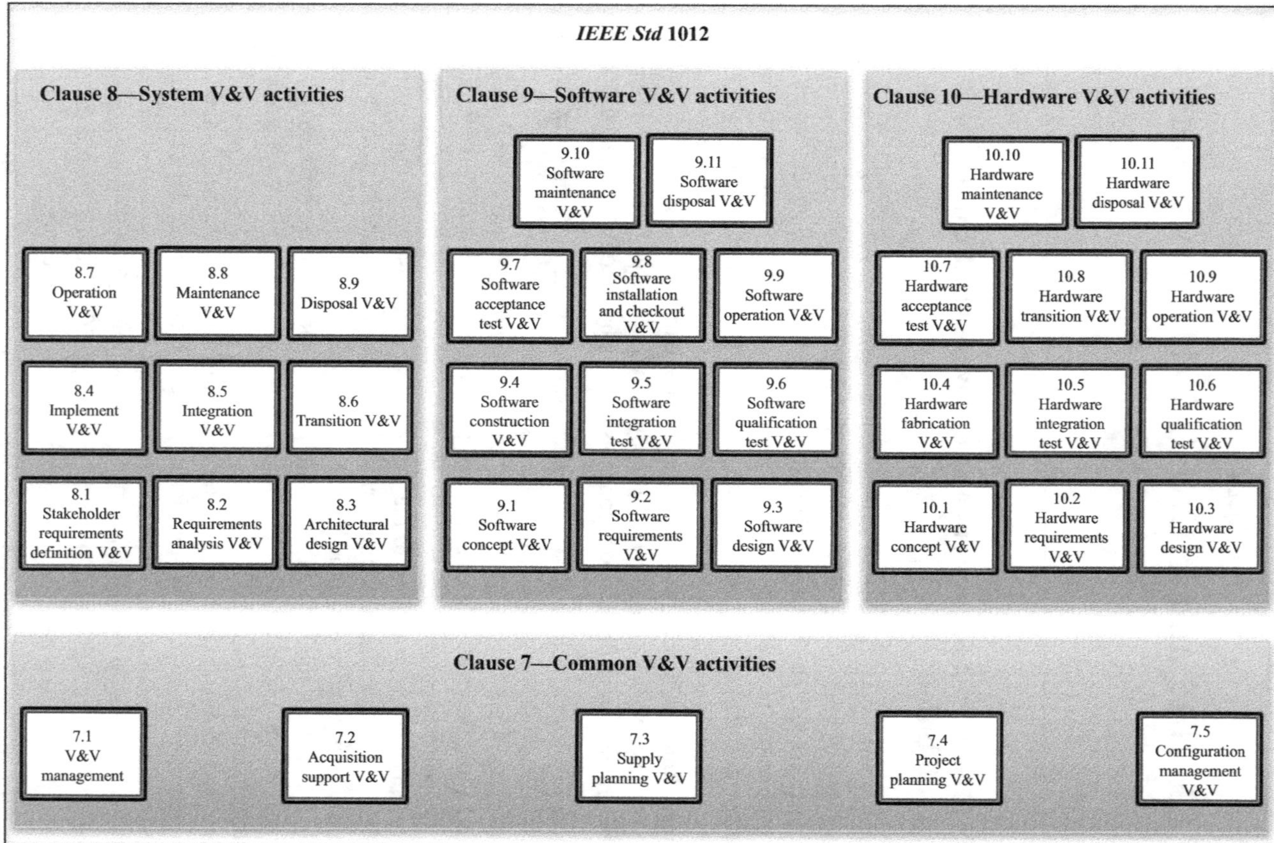

Figure 8.1 IEEE Std 1012

8.1.1 Integrity levels

The integrity level is related to the complexity of the system. It comprises the risk, criticality, safety and security needs, desired performance, expected reliability, and other quality characteristics. Integrity levels are used to determine the V&V tasks, activities, rigor, and level of intensity of the V&V to be performed.

The "Integrity-level Schema" established for the system, software, or hardware may be used to establish gradations of criticality and integrity to assure a complete partitioning of the potential effects from "no effect" to "worst case." The integrity levels established for a system element by the developer should result from agreements between the customer and the producer, and remain consistent with the regulatory requirements.

This standard was developed with a four-level schema to describe the minimum normative V&V requirements. Other integrity schemas can be defined ad hoc. For any integrity schema, the selected integrity levels must be mapped into the standard's schemas to demonstrate that the minimum V&V requirements are met. Because V&V is applied recursively from the system of interest down to each of the subsystems or elements at the next level, it is not necessary to use the same integrity level schema for all the subsystems or elements of the system of interest.

8.1.2 Common V&V activities

This standard defines several V&V activities common to any system, software or hardware V&V. Clause 7.1 (*V&V management*) concerns the V&V management activity, which monitors and evaluates all the artifacts produced during the V&V process. This activity involves: (1) a continual review of the effort, which depends on selected integrity level; (2) a revision of the process; (3) a coordination of the V&V activities and their results with other life cycle activities; (4) the performance of reviews and audits; and (5) the identification of process improvement opportunities.

Clauses 7.2 *acquisition support V&V* and 7.3 *supply planning V&V* concern the project initiation, negotiations, contract arrangements, supplier monitoring, acceptance, planning, execution and control, and review and evaluation, as well as delivery and completion activities.

Project planning V&V is the activity (Clause 7.4) that determines the scope of the project management and technical activities. It also identifies process outputs, project tasks, and deliverables and establishes schedules for project task conduct.

Finally, Clause 7.5 defines an activity (*configuration management V&V*) concerning the V&V configuration management.

8.1.3 Software V&V activities

Clause 9 of the standard defines 11 activities specific for software V&V:

Software concept V&V (Clause 9.1), which aims at verifying the system architectural design and system requirements analysis. The objective of software concept V&V is to verify the allocation of system requirements and validate the selected solution. The tasks of the activity are: (1) concept documentation evaluation;

(2) hardware/software/user requirements allocation analysis; (3) traceability analysis; (4) criticality analysis; (5) hazard analysis; (6) security analysis; and (7) risk analysis.

Software requirements V&V (Clause 9.2), which aims at assuring the correctness, completeness, accuracy, testability, and consistency of the system software requirements. The tasks of the activity are (1) requirements evaluation; (2) interface analysis; (3) traceability analysis; (4) criticality analysis; (5) software qualification test plan V&V; (6) software acceptance test plan V&V; (7) hazard analysis; (8) security analysis; and (9) risk analysis.

Software design V&V (Clause 9.3), which aims at demonstrating that the transformation of the software requirements into an architecture and a detailed design for each software component is correct, accurate, and complete with respect to the software requirements and that no unintended features are introduced. The tasks of the activity are (1) design evaluation; (2) interface analysis; (3) traceability analysis; (4) criticality analysis; (5) software component test plan V&V; (6) software integration test plan V&V; (7) software component test design V&V; (8) software integration test design V&V; (9) software qualification test design V&V; (10) software acceptance test design V&V; (11) hazard analysis; (12) security analysis; and (13) risk analysis.

Software construction V&V (Clause 9.4), which aims at verifying and validating that the transformations of models into code are correct, accurate, and complete. The tasks of the activity are (1) source code and source code documentation evaluation; (2) interface analysis; (3) traceability analysis; (4) criticality analysis; (5) software component test case V&V; (6) software integration test case V&V; (7) software qualification test case V&V; (8) software acceptance test case V&V; (9) software component test procedure V&V; (10) software integration test procedure V&V; (11) software qualification test procedure V&V; (12) software component test execution V&V; (13) hazard analysis; (14) security analysis; and (15) risk analysis.

Software integration test V&V (Clause 9.5), which aims at assuring that the software requirements and system requirements allocated to the software are validated as each software component (e.g., unit) is incrementally integrated. The tasks of the activity are (1) software integration test execution V&V; (2) traceability analysis; (3) hazard analysis; (4) security analysis; and (5) risk analysis.

Software qualification test V&V (Clause 9.6), which aims at demonstrating that the integrated software product satisfies its requirements. Software qualification (e.g., testing, analysis, or inspection) is performed on the complete software element. The tasks of the activity are (1) software qualification test execution V&V; (2) traceability analysis; (3) hazard analysis; (4) security analysis; and (5) risk analysis.

Software acceptance test V&V (Clause 9.7), which aims at assuring that the software satisfies its acceptance criteria and at enabling the customer to determine whether or not to accept the integrated software product. The tasks of the activity are (1) software acceptance test procedure V&V; (2) software acceptance test execution V&V; (3) traceability analysis; (4) hazard analysis; (5) security analysis; and (6) risk analysis.

Software installation and checkout V&V (Clause 9.8), which aims at demonstrating the correctness of the software installation in the target environment. The tasks of the activity are (1) installation configuration audit; (2) installation checkout; (3) hazard analysis; (4) security analysis; and (5) risk analysis.

Software operation V&V (Clause 9.9) evaluates new constraints in the system, assesses the proposed system changes and their impact on the software, and evaluates the operating procedures for correctness and usability. The tasks of the action are (1) evaluation of new constraints; (2) operating procedures evaluation; (3) hazard analysis; (4) security analysis; and (5) risk analysis.

Software maintenance V&V (Clause 9.10), which assesses the proposed software system changes and their impact on the software, evaluates the anomalies that are discovered during operation, assesses the migration requirements, assesses the retirement requirements, and reperforms the V&V tasks. The proposed changes are assessed by the proposed/baseline change assessment task of the management of V&V activity. The tasks of the activity are (1) VVP revision; (2) anomaly evaluation; (3) criticality analysis; (4) migration assessment; (5) retirement assessment; (6) hazard analysis; (7) security analysis; (8) risk analysis; and (9) task iteration.

Software disposal V&V (Clause 9.11), which evaluates the result of the software disposal process.

8.2 ISO/IEC 29119 software testing

ISO/IEC/IEEE 29119 software testing is a set of standards concerning software testing. It can be used within any software development life cycle or organization. There are currently five standards:

- ISO/IEC 29119-1: concepts & definitions
- ISO/IEC 29119-2: test processes
- ISO/IEC 29119-3: test documentation
- ISO/IEC 29119-4: test techniques
- ISO/IEC 29119-5: keyword-driven testing

The ISO/IEC/IEEE 29119 standards replace, evolve, or integrate other existing software testing standards, such as IBS 7925-1 vocabulary of terms in software testing, BS 7925-2 software component testing standard, EEE 829 test documentation, and IEEE 1008 unit testing.

Figure 8.2 shows the underlying model, which is the basis for the new set of standards and comprises five entities. The test processes constitute the central core of the model, defined in part 2 of the standard. The test documentation (part 3) is produced during the execution of the processes. Thus, the test documentation describes the outputs of the test processes. The design of the test cases, as part of the processes, exploits the testing techniques that are defined separately in part 4 of the standard. The test assessment is performed using a process assessment model based on the process reference model defined in part 2. The terminology used by the other parts of the model is defined in the vocabulary (part 1).

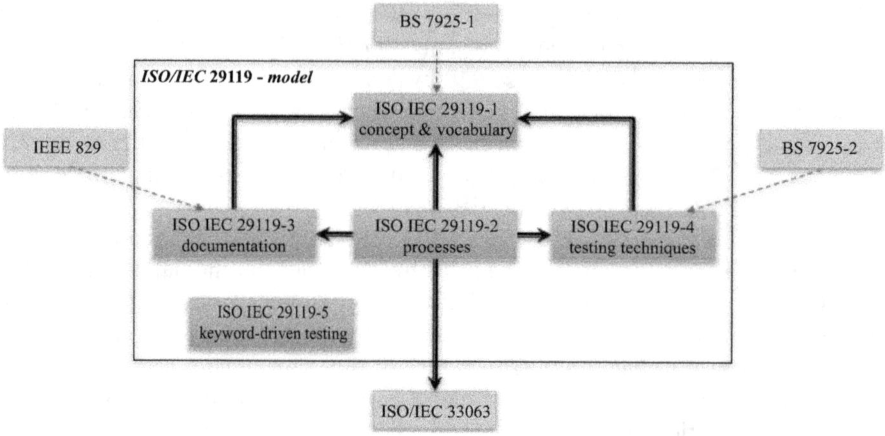

Figure 8.2 ISO IEC 29119 model

8.2.1 ISO/IEC 29119-1: concepts & definitions

The aim of ISO/IEC/IEEE 29119-1 is to provide an introduction to the set of standards and it includes the overall set of definitions on which all other parts are built.

This part explains what the different parts of the standard include and how the standard can be used by those adopting different life cycle models including, for example, the V-model and Agile approaches.

The most relevant sections of this part are the following:

- Introduction to software testing
- Software testing in an organizational and project context
- Generic testing processes in the software life cycle
- Risk-based testing
- Test subprocesses
- Test practices
- Automation in testing
- Defect management
- The role of testing in V&V
- Metrics and measures

8.2.2 ISO/IEC 29119-2: test processes

ISO/IEC/IEEE 29119-2 specifies a generic process model for software testing, which can be adopted within different software development life cycles. The model defines test processes that can be used to govern, manage, and implement software testing in any organization, project, or testing activity.

The processes are defined using a three-layer model. The model includes an organizational level, a test management of individual test phases level and a dynamic

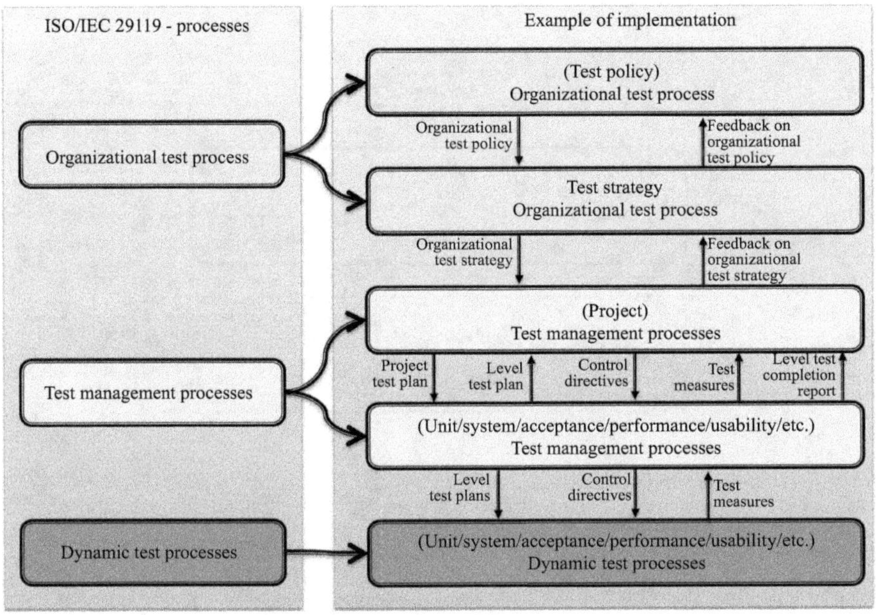

Figure 8.3 Three-layer architecture for ISO/IEC 29119 processes

testing level. The standard still lacks a static testing process, which should be included in the future.

Figure 8.3 provides an example of how these layers may be implemented in a relatively large and mature organization; that is, an organization having both an organizational test policy and an organizational test strategy. The figure shows how the generic processes of part 2 of the standard may be instantiated for the application at different levels. The organizational test process is instantiated to develop and maintain both the organizational test policy and the organizational test strategy. The test management processes enable the development and implementation of the project test plan. They are also used for each subsequent phase or type of testing, for each phase of which a separate test plan is created. All test plans within this standard are expected to include a consideration of both static and dynamic testing. The lowest layer is currently dedicated to dynamic testing, which is implemented whenever the test plan requires testing for individual software units, the entire system, the performance assessment or whenever there is a requirement for dynamic testing.

Figure 8.4(a) shows the complete set of eight test processes defined by the standard. It is worth noting that there is just one process defined at the organizational level. Each process comprises a set of activities, and each of these activities is composed of a set of specific tasks. There is only one organizational test process and, as can be seen, it comprises three activities. This process is used for the development and maintenance of the organizational level test specifications, which include the test policy

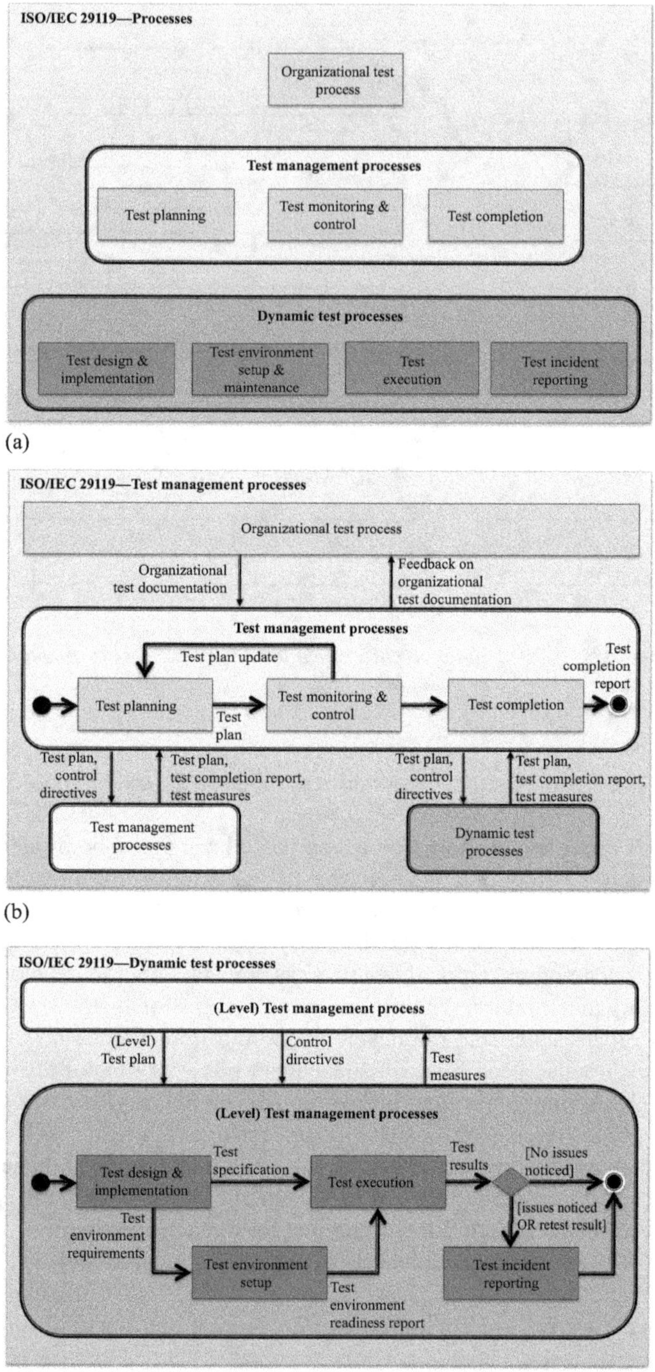

Figure 8.4 All ISO/IEC 29119 processes: (a) all processes, (b) test management processes, and (c) dynamic test processes

and organizational test strategy. The organizational level test specifications typically apply across a number (if not all) of the projects in the organization and therefore they are not project-specific. The aim is to ensure that a consistent approach is maintained across all projects. If the organization is very large, it may be the case that similar projects are arranged in homogeneous classes, and there may be organizational test specifications that are specific to each class.

The test policy defines the scope of the testing activity and describes the principles and high-level objectives expected of all testing for all the projects within the organization. It is supposed to be a brief-high-level document, understandable by executives and aligned with the quality policy. The test policy provides guidance about the constraints on the organizational test strategy and acts as a framework within which all the testing activities should be performed. A test policy could, for example, establish that all testing documentation must comply with ISO 29119-3. The organizational test strategy, instead, is a more technical document that concerns the expected practice of testing within the organization. It must conform to the test policy by defining a set of reusable testing guidelines that will represent the basis for specific test plans and reduce the decision-making ensuring consistency across projects. A piece of an organizational test strategy could be that the exit criteria for unit testing must be greater than 60% with respect to the decision coverage.

There are three test management processes used to manage lower level testing processes for each project, namely dynamic testing, static testing (not shown because they are not in the standard) and lower level test management, when necessary. If the project is large, then the project test plan, at the project level, may require further test plans to be generated and implemented for individual test phases such as, for example, the system testing test plan or acceptance testing test plan.

All the constraints and guidelines provided by the organizational test policy or the organizational test strategy should apply while these test management processes are performed. In any case in which a deviation is necessary, it must be supported by good reasons and documented.

The test management processes issue the test plan, which is used as the basis of the actual testing performed. Several activities are carried out to generate the test plan. One of the most important is the design of the test strategy for the project. The test strategy may affect decisions such as which features are to be tested, which test techniques are to be exploited and how test completion criteria are defined. This test strategy may be quite different from the organizational test strategy. Indeed, the organizational test strategy provides generic guidelines on how to test across many projects. On the contrary, the test strategy at this stage forms part of the test plan and defines the specific testing to be conducted on the current project. It is worth noting however that even if deviations from the organizational test strategy may be needed, the project test strategy should normally be aligned with the guidelines and constraints defined within the organizational test strategy. As a consequence, there should necessarily be a good degree of similarity between these two strategies.

The test monitoring and control comprises four activities driven by the test plan objectives. One challenge is to identify test measures which can be used to determine how closely the actual testing is following the plan. The progress of the testing (not shown in Figure 8.4(b)) and the feedback from this process, which includes updates to the test plan, are reported.

Moving down a level, Figure 8.4(c) shows the dynamic test processes. It should be noted that the test environment setup and test incident reporting could be shared between the static and dynamic testing processes. The test plan, the organizational test strategy, and the test policy could all concur to drive the dynamic test activities. For example, the test plan could specify what features are to be tested, the organizational test strategy could specify that certain test completion criteria are to be achieved and the test policy could specify that the incident reporting is to follow a defined organization-wide instant management process.

8.2.3 ISO/IEC 29119-3: test documentation

Part 3—test documentation—is closely related to part 2 (test processes) because the outputs of the process defined in part 2 generally correspond to the test documentation defined in part 3.

The test documentation defined by ISO 29119-3 is based on IEEE Std 829, which is now superseded. It includes three major classes of documents: Organizational Test Process Documentation, Test Management Process Documentation and Dynamic Test Process Documentation. The first class includes the Test Policy and Organizational Test Strategy. The second comprises the test plan (including a test strategy), test status report and test completion report. The third includes the following document types: Test Design Specification, Test Case Specification, Test Procedure Specification, Test Data Requirements, Test Data Readiness Report, Test Environment Requirements, Test Environment Readiness Report, Actual Results, Test Result, Test Execution Log, and Test Incident Report.

8.2.4 ISO/IEC 29119-4: test techniques

ISO 29119 Part-4 defines test case design techniques along with test completion criteria and corresponding coverage measures.

Test case design techniques are formally defined in the normative part of the standard and some examples showing how the techniques could be applied are provided in the annexes.

The basis for this standard is the BS-7925-2 Component Testing standard, which was superseded by ISO/IEC/IEEE 29119-4. Some appreciable differences regard non-functional testing such as that for quality characteristics. ISO 29119-4 includes eleven black-box techniques, six white-box techniques, and one experience-based technique:

- Specification-based testing techniques:
  - Equivalence partitioning
  - Classification tree method
  - Boundary value analysis
  - State transition testing
  - Decision table testing
  - Cause–effect graphing

  - Syntax testing
  - Combinatorial test techniques
  - Scenario testing
  - Random testing
- Structure-based testing techniques:
  - Statement testing
  - Branch testing
  - Decision testing
  - Condition testing
  - Data flow testing
- Experience-based testing techniques:
  - Error guessing

Part 4 also provides informative definitions of a variety of quality-related types of testing and provides examples of how the test case design techniques in the lists above can be applied within the following types of testing:

- Accessibility testing
- Back-up/recovery testing
- Compatibility testing
- Conversion testing
- Disaster recovery testing
- Functional testing
- Installability testing
- Interoperability testing
- Localization testing
- Maintainability testing
- Performance-related testing (e.g., load testing, stress testing, and endurance testing)
- Portability testing
- Procedure testing
- Reliability testing
- Security testing
- Stability testing
- Usability testing

8.2.5 ISO/IEC 29119-5: keyword-driven testing

ISO/IEC/IEEE 29119-5 concerns the definition of a standard to support the keyword-driven testing. Such a kind of testing provides a way of describing test cases by means of keywords, which are names associated with actions required to perform a specific step in a test case. The advantage of using keywords instead of natural language to describe such steps is that the resulting test cases can be more easily understood and automated.

Part IV

Verification and validation techniques

Chapter 9
Static testing

9.1 Introduction and background

Verification is the activity relating to the evaluation of the software system in order to determine whether the product of a given development phase satisfies the requirements established before the start of that phase. The focus is on building the product correctly.

Validation, instead, concerns the evaluation of the software system in order to determine whether the product meets its intended use. The focus is on building the correct product.

Testing is one of the best tools to verify and validate a software system.

Before going into detail about verification, validation, and testing, however, it is worth clarifying some of the terminology that often causes confusion: an **Error** [or **Mistake**], is a human action that produces an incorrect result; a **Fault** (or **Defect**, or **Bug**) is the manifestation of an error in the system; a **Failure** is a deviation of the behavior of the system from its expected delivery or service, that is the inability of the system or one of its components to perform its required function within the specified performance requirement.

In other words, a failure is a deviation of the system from its intended (expected) purpose and a fault is a defect (or bug) of the software that may potentially cause failures. It is important to note that the presence of faults does not always cause failures. Indeed, a fault remains inactive until the defective part of the system is solicited during the execution. Even if a fault is activated due to the execution of the defective part of the system, it may be masked by other internal conditions of the system and no failure may be observed.

Testing is then the activity that aims at stressing the system and causing failures. Testing is also a means to measure system quality and build confidence. High-confidence systems are robust (the manage failures) and can justifiably be trusted, especially when used in life-, safety-, security, and mission-critical situations. High confidence is exactly what the manufacturer needs when realizing a software system for medical purposes.

Several testing techniques exist. These are categorized as either static or dynamic. Static testing is performed without the execution of the software and generally may be applied to all kinds of document. Specific static testing techniques have been developed to test source code. One of the most important values provided by static testing consists in the possibility of anticipating dynamic testing and defect detection in early development phases with a dramatic reduction in fixing costs.

9.2 Static testing

Static testing includes several approaches depending on the "object" to be tested. Examinations performed by human beings can be conducted in order to check the quality level of any kind of project document. Such an examination may be informal or formal.

Among the informal examinations, there are **proof-reading** and **informal reviews**. They are substantially based on the simple idea that reviewing a document may lead to the discovery of defects. Such techniques differ from one another because proof-reading is typically performed by the author of the document himself, whereas informal reviews are performed by another person (e.g., a colleague) who can analyze the artifact from a "different point of view."

More formal static testing techniques for generic documents are **formal reviews**, **walkthroughs**, and **inspections**. They share the idea that reviews are planned structured activities and reviewers are formally in charge of analyzing the artifact and issuing a written feedback used to improve the quality.

Walkthroughs and inspections are the most formal types of group review that take place in a meeting where the author illustrates the document to be reviewed. In a walkthrough, the author guides the group during the presentation of the document, so all the participants have the same understanding of the artifacts. Consensus on changes must be achieved. In the case of an inspection, the reviewers are provided with the document to be reviewed in advance, so all the participants arrive at the meeting with their own ideas.

Both approaches define roles:

- the author, who produces the document under review/inspection;
- the leader/moderator, who plans the review/inspection, chooses the participants, conducts the meeting, performs any follow-up, and manages the metrics;
- the reviewers/inspectors, who are specialized in finding faults;
- the manager, the project manager (excluded from some types of review), who needs to plan the timing and resources for the reviews/inspections;
- the recorder, who takes notes during some types of review meeting.

A typical inspection process includes the following phases:

1. Planning: this concerns the definition of the criteria for the review, the selection of the participants, the allocation of roles, the definition of the input and output criteria, the choice of which parts of the document to review;
2. Kick-off: this takes place typically with a meeting to (1) check the input criteria, (2) distribute the documents, and (3) explain the processes and objectives of the documents to the participants;

3. Individual preparation: this is preliminary to the review meetings and enables the reviewers individually to gain an awareness of the content of the documents;
4. Individual review: each reviewer checks the documents to identify potential defects and prepares questions to ask and comments to make during the review meeting;
5. Review meetings: these are focused on the review results which are documented with minutes or reports. During the meeting, notes may be made and recommendations for defect management issued. Moreover, decisions about the treatment of defects are taken with particular regard to any need for reworking;
6. Reworking: this aims at the correction of defects reported during the review meetings. Reworking is typically performed by the author;
7. Follow-up: this is the concluding activity during which the defects that have been corrected are checked. Metrics data may be collected and exit criteria verified.

The benefits of reviews include the early detection and correction of defects, an improvement in the productivity of development and a consequent reduction in time, a reduction in costs and testing times, a reduction in the overall costs of the life cycle, a reduction in the number of defects and an improvement in the communication. It has been demonstrated that such techniques reduce the number of faults by a factor of 10.

Reviews may result in the discovery of omissions, for example in the requirements, which are rarely found in dynamic testing. Static testing and dynamic testing are complementary, not substitutive of one another: different techniques can result in the discovery of different types of defects effectively and efficiently. Compared to dynamic testing, static techniques can reveal causes of potential failures (defects) rather than the failures themselves. Typical defects that are easier to find in reviews than in dynamic testing are deviations from standards, requirements defects, design defects, and incorrect interface specifications.

9.3 Static analysis

Static analysis is a set of techniques relating to the static testing of software code. Unlike other types of documents, software code is partially analyzed by automatic instruments (programs) which are nowadays running in Integrated Development Environments.

Static analysis techniques exploit different representations of the structure and properties of a software unit: **control flow graphs**, **data dependence graphs**, and **control dependence graphs**.

9.3.1 Control flow analysis

A control flow graph represents the flow of execution of the instructions of the software unit under all possible conditions. In the graph, a node represents a sequence of instructions executed unconditionally decision or up to a point of simple decision.

Edges connect two sequences of instructions. Each point of simple decision in a program has one incoming edge and a pair of outgoing edges.

An example of a software unit (in this case a Java function) and its control flow graph are reported respectively in Listing 9.1 and Figure 9.1. It is worth noting that the first five instructions starting from the top could be included in one node in accordance with the previous definitions. However, the use of distinct nodes for each instruction may facilitate the control and data dependence analysis as shown in the next sections.

Listing 9.1 A software unit

```
private int ZC(int[] vet){
        // This function calculates
        // the Zero Crossing
    int count = 0;
    int iteractions = samples -1;
    int prod;

    for(int i = 0; i<iteractions; i++){
        if((vet[i]*vet[i+1] == 0) && i>0)
            prod = vet[i-1]*vet[i+1];
        else
            prod = vet[i]*vet[i+1];

        if(prod < 0)
            count++;
    }

    return count;
}
```

Figure 9.1 shows an intraprocedural control flow graph. In order to represent extra-procedural control flow, it is possible to realize similar graphs, called **call graphs**, where nodes represent procedures (e.g., object methods, c functions) and edges represent function calls.

The control flow diagram allows the acquisition of a quantitative measure of the software unit complexity, namely the **cyclomatic complexity**. The cyclomatic complexity of a section of source code is the number of linearly independent paths within it. It is given by the number of simple decision nodes (in the graph) plus one. The cyclomatic complexity of the example listed in 9.1 is four.

Thomas McCabe, who defined cyclomatic complexity, reported remarkable results in his paper [34]. He presented several examples of modules that were poorly structured having cyclomatic complexity values ranging from 16 to 64 and argued that such modules tended to be quite buggy. Then he gave examples of developers

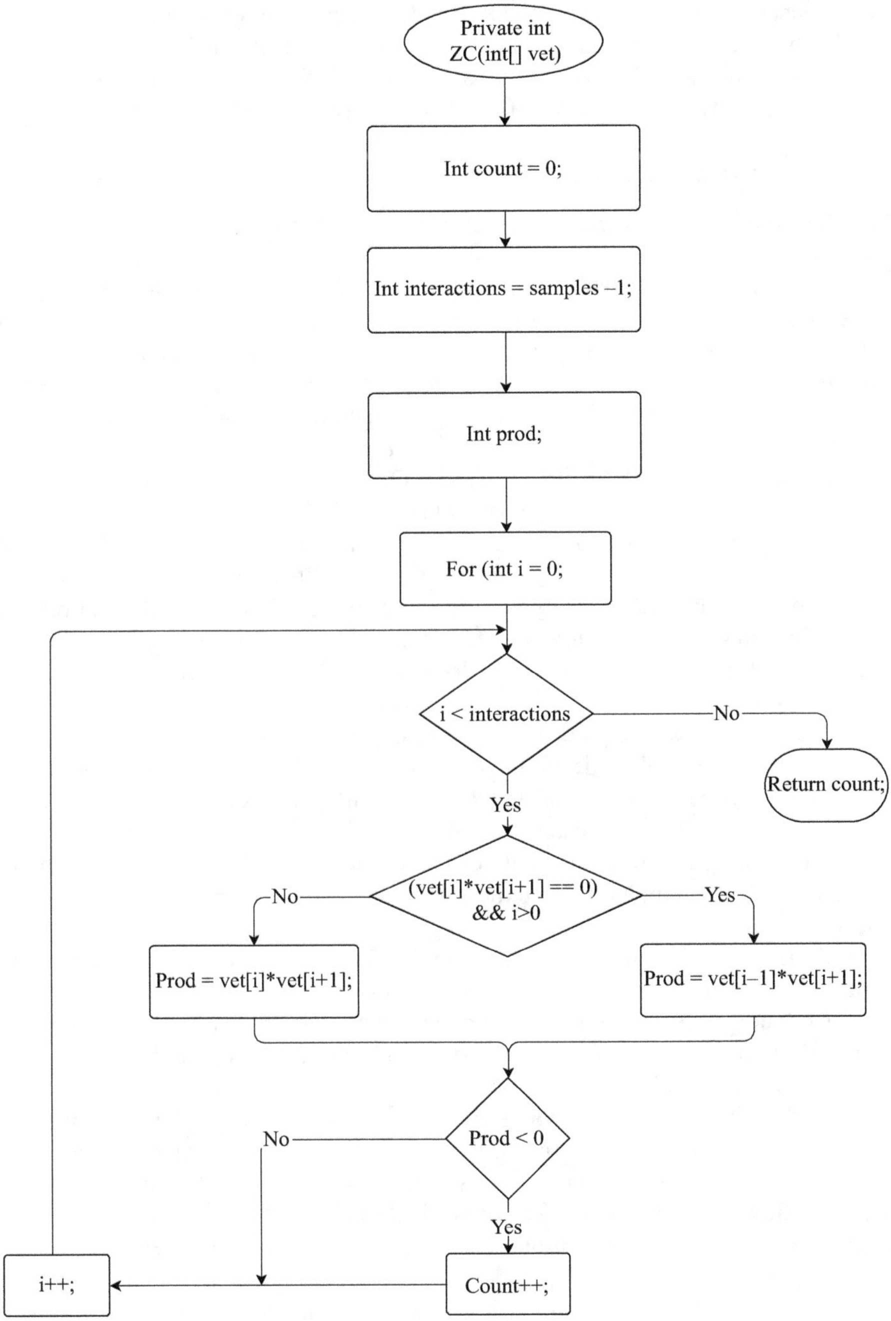

Figure 9.1 Control flow graph for the code listed in 9.1

who consistently wrote low complexity modules (in the three to seven cyclomatic complexity range) who regularly achieved a far lower defect density.

A control flow graph is an essential tool in the design of test cases for dynamic testing, but it is useful also for the realization of both data and control dependence analysis.

9.3.2 Data dependence analysis

Data flow models (e.g. Figure 9.2) complement models of the control flow of a program.

Testers often need to reason about information flow, or how a value at one point in the program depends on values of the same or different variables at some prior point in execution. For example, natural questions to ask about a variable *x* at some point in program execution are "where did this value come from?" and "where could it be used?" Data flow models capture this dependence information naturally.

An example of using a data flow model is the building of an automatic analysis system to detect vulnerability to Structured Query Language (SQL) injection" or "command injection" attacks. If the contents of a web form are later used in forming a database query, we need to be sure that there is no way for an attacker to use malformed SQL to obtain privileged information or destroy the database. A program analysis to detect this vulnerability must use information about how information that is input from the web form is checked, transformed, and finally used in the query; in other words, it must be based on a data flow model of the application.

A **def–use pair** is the most basic data flow relationship. A value is "defined" (produced or stored) at one point and "used" (referenced or retrieved from a variable) at another point. Formally, a definition (def) is a point where a variable is written: (1) the variable is declared and initialized, (2) the result of an expression is assigned to the variable, or (3) the variable gets the value of an incoming parameter in a function call. A use is a point where the value of the variable is read: (1) in expressions of assignment, (2) in a conditional statement, (3) in an outgoing parameter in a function call, or (4) in a return of a function.

A **definition-clear path** is a path along the control flow graph from a definition to a use of the same variable without another definition of the variable between. In the case of another definition existing along the path, then the latter definition would kill the former. A def–use pair is formed if, and only if, there is a definition-clear path between the definition and the use.

Data flow information is associated with locations in the program or, equivalently, with nodes in the program control flow graph. It is possible to mark each location with the sets of variables defined and used at those locations. Note that in node A (the beginning of the procedure) vector *vet* is defined by the values passed through parameters, whereas the declaration of *prod* is treated as an assignment of a special value "uninitialized."

Lines G and H (nodes G and H in the graph below) show how an assignment statement typically defines one variable (prod, in this case) and uses several variables.

Line M shows that returning a value from a method or function is also a use of that value.

Listing 9.2 Statement (Node) labeling and def–use points

```
private int ZC(int[] vet){                              // A: Def vet[]
        // This function calculates
        // the Zero Crossing
        int count = 0;                                  // B: Def count
        int iteractions = samples -1;                   // C: Def iteractions;
                                                        //     Use samples
        int prod;                                       // D: Def prod

        for(int i = 0; i<iteractions; i++){             // E: Def i;
                                                        //     Use i, iteractions
                if((vet[i]*vet[i+1] == 0) && i>0)       // F: Use i, vet
                        prod = vet[i-1]*vet[i+1];       // G: Def prod; Use i, vet
                else
                        prod = vet[i]*vet[i+1];         // H: Def prod; Use i, vet

                if(prod < 0)                            // I: Use prod
                        count++;                        // L: Def count; Use count
        }

        return count;                                   // M: Use count
}
```

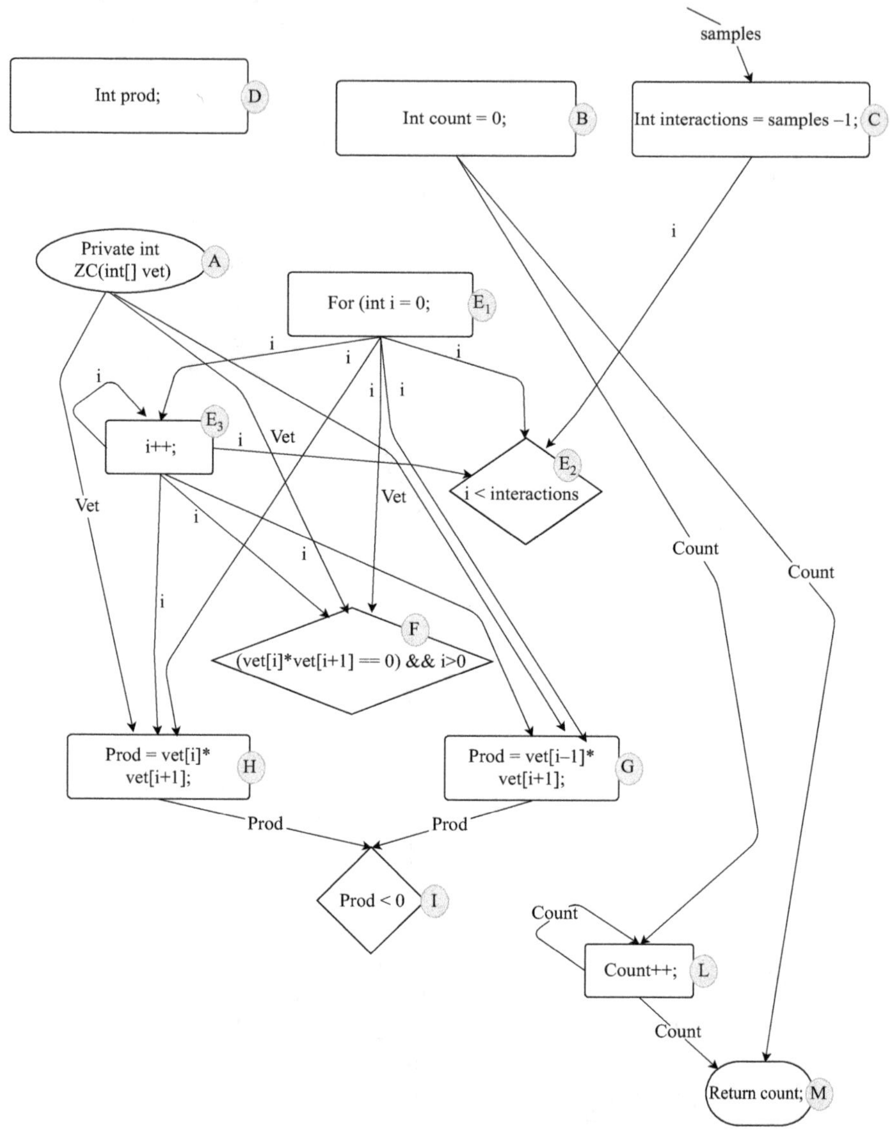

Figure 9.2 Data dependence graph for the code listed in 9.2

Some preliminary analysis may be easily conducted. Indeed, (1) a definition without any successive use is an anomaly in the program; (2) a use without any previous definition may be a fault (this is the case of line C of code listed in 9.2 unless the variable *samples* is an attribute of the class); and (3) any killing definition is also an anomaly.

Table 9.1 Combinations of atomic actions: def, use, kill

Combination			Explanation
1.	~d	First define	Allowed
2.	du	Def–use	Allowed
3.	dk	Def–kill	Potential bug. Data were never used
4.	~u	first use	Potential bug. Data were used without definition. It may be a global variable or an object attribute, defined outside the routine
5.	ud	Use–def	Allowed. Data used and then redefined
6.	uk	Use–kill	Allowed
7.	~k	First kill	Potential bug. Data are killed before definition
8.	ku	Kill–use	Serious defect. Data are used after being killed
9.	kd	Kill–def	Usually allowed. Data are killed and then redefined. Some theorists believe this should be disallowed
10.	dd	Def–def	Potential bug. Double definition
11.	uu	Use–use	Allowed
12.	kk	Kill–kill	Likely bug
13.	d~	Define last	Potential dead variable. May be a global variable used in another context
14.	u~	Use last	Allowed. Variable was used in this routine but not killed off
15.	k~	Kill last	Allowed. Normal case

Rex Black and Jamie L. Mitchell reported in Table 9.1 the analysis of 15 potential combinations of the atomic actions Def, Use, and Kill [35]. They commented on such combinations as follows:

1. ~d, this is the normal way a variable is originated: it is declared. However, a declaration may be not enough for a definition. Indeed, a variable is declared to allocate space in memory for it; at that point, however, the value the variable holds may be unknown (although some languages have a default value that is assigned, often zero or NULL). The variable needs to be set before it is used.
2. du, this is the normal way a variable is used. Defined first and then used in an assignment or decision predicate.
3. dk, this is likely to be a software defect since the variable was defined and then killed off. The question must be asked as to why it was defined. Is it dead? Or was there a thought of using it but the wrong variable was actually used in later code? If creating the variable caused a desirable side effect to occur, this might have been done purposefully.
4. ~u, this is a potential defect since the data were used without a known definition. It may not be a bug because the definition may have occurred in a different scope (e.g., a global variable or an object attribute). This should attract attention and be investigated by the tester. Indeed, race conditions may mean that the variable never gets initialized in some cases. Moreover, global variables should not be used except for a few very special purposes.

5. ud, this is allowed where the data are read and then set to a different value. This can all occur in the same statement where the use is on the right side of an assignment and the define is the left side of the statement, such as $i = i + 1$.
6. uk, this is a normal behavior.
7. $_\sim$k, this is a potential bug where the variable is killed before it is defined. It may simply be dead code, or it might be a serious failure waiting to happen, such as when a dynamically allocated variable is destroyed before actually being created, which would cause a runtime error.
8. ku, or kill-use. This is always a serious defect where the programmer has tried to use the variable after it has gone out of scope or been destroyed. For static variables, the compiler will normally catch this defect; for dynamically allocated variables, it may cause a serious runtime error.
9. kd, this is usually allowed as long as a value is destroyed and then redefined. Some theorists believe this should be disallowed; once a variable is destroyed, bringing it back is asking for trouble. Others do not believe it should be an issue. Testers should evaluate this one carefully if it is allowed.
10. dd, this is a potential bug. The variable is defined, then redefined without any previous use. It looks very strange and may imply a logic mistake or some dead code.
11. uu, this is a normal behavior.
12. kk, this is likely to be a bug, especially when using dynamically created data. Once a variable is killed, trying to access it again, even to kill it, will cause a runtime error.
13. d$_\sim$, this is a potential bug, as in the case of a dead variable that is never used. However, it might also be that the variable is meant to be global and used elsewhere. All global variables require accurate analysis.
14. u$_\sim$, this is a possible behavior. The variable is used at last and never explicitly killed.
15. k$_\sim$, this is the normal case. The variable is killed at last.

More detailed data flow analysis may be conducted as well. **Reaching definition analysis** statically determines which definitions may reach a given point in the code, that is it aims at providing answers to questions like "Which definition defines the value used in statement L (*count++;*)?" In this specific case, the value used for the variable *count* in the assignment comes from line B the first time the cycle is executed, otherwise from line L itself. More formally, a definition *d* reaches a line *p* if there exists a path from the point immediately following *d* to *p* such that *d* is not killed along that path.

Another sort of data flow analysis is **live variable analysis** that aims at determining if each variable is "live" in each basic block. A variable *v* is live at point *p* if its value is used in at least one path in the flow graph starting at *p*. Otherwise, the variable is dead. This analysis may be performed by automatic tools just to identify and remove dead code or variables.

Other data flow analysis techniques are **available expression**, **constant propagation**, and **definite assignment analysis**. In general, data flow analysis may be

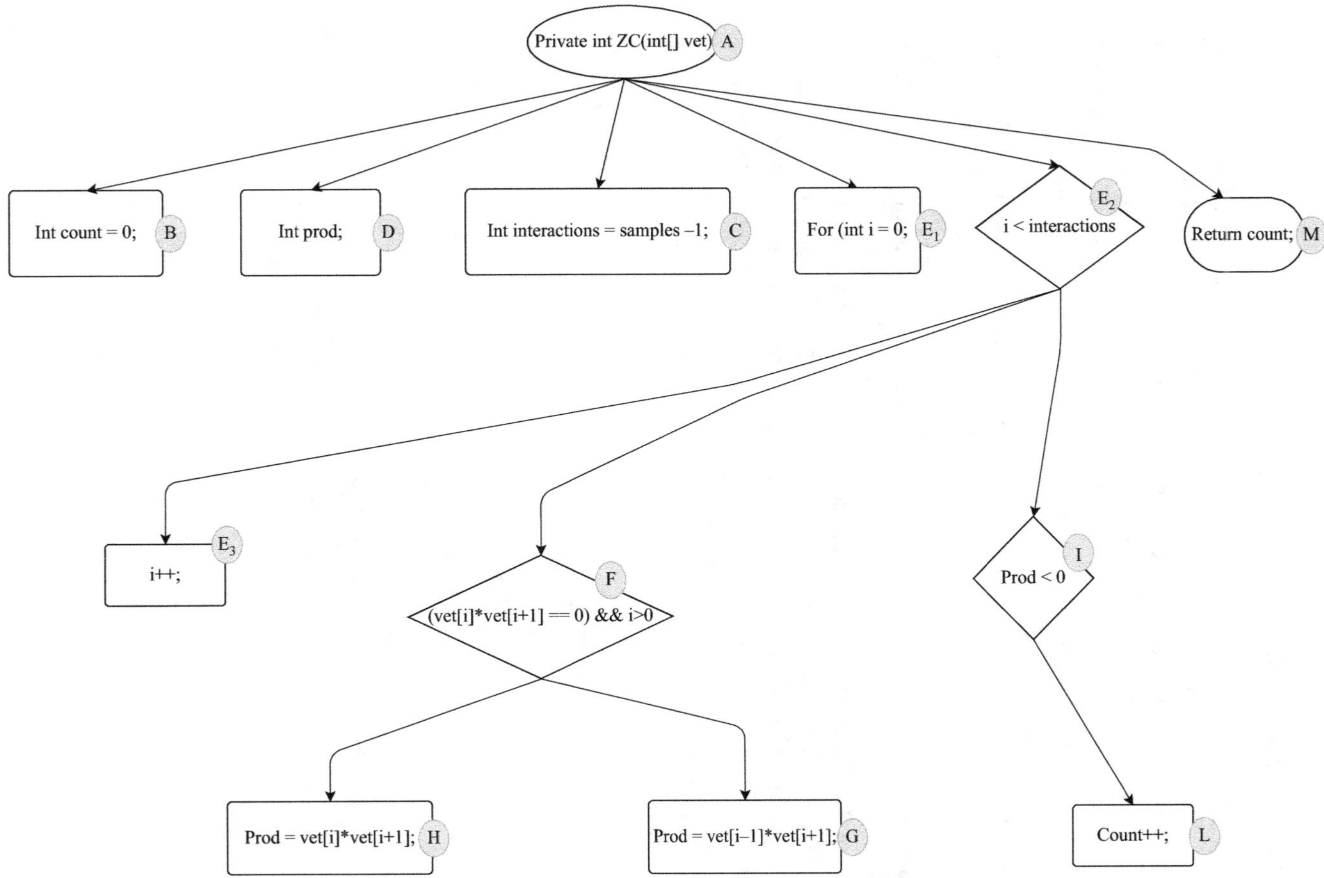

Figure 9.3 Control dependence graph for the code listed in 9.2

applied for several reasons, including the symbolic debugging of code and static error checking.

9.3.3 Control dependence analysis

Dependence analysis often involves both the flow of information (data flow) and flow of control. The basic idea behind the concept of control dependence is that a node controls another node if it determines whether that node executes. For example, an "if-else" statement determines the execution of two alternative paths depending on the result of a Boolean expression. An instruction of the "else" path is executed if and only if the computed Boolean expression is false; thus, the instruction is dominated by this "if" statement.

Formally, in a directed graph, node M dominates node N, if every path from the root to N passes through M. A node may have many dominators, but there is a unique immediate dominator of node N that is the closest to N on any path from the root. The immediate dominator relation forms a tree. Indeed, each node has a unique immediate dominator.

Figure 9.3 shows the control dependence graph for the code listed in 9.2. It is obtained starting from exactly the same (labeled) nodes of the control flow diagram and adding edges to represent immediate domination relations; an edges from node M to node N means that M is the immediate dominator of N, that is M is the last node that can make the execution of N avoidable.

One remarkable use of a control dependence graph is for the elimination of dead code.

Chapter 10
Dynamic testing

10.1 Introduction

Dynamic testing (or dynamic analysis) aims at verifying the dynamic behavior of a system. It encompasses the design and execution of test cases. Although the design of a test case can already be performed in the early stages of the development of the software system (e.g., the acceptance tests), the execution, and thus the verification of the behavior, can, of course, be performed only after the system has at least partially been realized. In other words, dynamic testing needs the software to have been compiled and executed with some specific input values and internal conditions in order to compare the obtained results with those expected.

According to the ISO/IEC 29119 standard, three main dynamic test techniques exist: specification-based, structure-based, and error-guessing. **Specification-based** (also called "Black-Box") techniques allow testers to derive test cases from the analysis of the "Test Basis," which is the set of specifications defined at various stages of the development cycle. Specifications, either functional or nonfunctional, drive the test design without reference to the internal structure of the system.

Structural-based (also called "White-Box") testing relies on models of the internal structure of the system to derive dynamic test cases. Structural-based testing is necessary because if only specification-based techniques are applied, a large amount of code might never be executed. A well-designed system may contain a lot of code in order to handle the exceptional conditions that may occur under unexpected inputs. Developers realize the system in such a way as to be resilient to these unlikely conditions. As a result, levels of complexity are piled on top of levels of complexity, and the system is hard to test in depth. The possibility of looking insight the system, examining its internal structure and defining test cases that solicit specific paths of execution or blocks of code, allows us to achieve a higher level of confidence. White-box testing is also useful for a measurement of the test coverage; i.e., a measurement of the extend of the system exercised.

Error guessing is based on the tester's experience. She/he makes "guesses" about mistakes that programers might make while solving specific problems and then develops tests for them. This approach is also known as "Gray-Box" testing because it requires the tester to have some idea about internal algorithms, possible programing mistakes, and consequent software bugs, about how these bugs manifest themselves as failures, and about how to test the system to make such failures take place.

In the following sections, we are going to discuss how to design, create, and execute dynamic testing.

10.2 Specification-based testing technique

10.2.1 Equivalence partitioning

Equivalence partitioning exploits a model of the unit or the system that classifies the input and output data in "equivalent partitions." An equivalent partition is a set or group of values, for given input or output data, for which the unit or the system is supposed to have the same behavior. Both valid and invalid value partitions are derived from the specification of the component's behavior.

Test cases must be designed to exercise partitions. The goal is to test the system against at least one value of each valid input/output partition and one value of each invalid input partition.

To show an application of this technique, we will consider the simple case of a web interface for the registration of a new product to sell on an eCommerce website. We will assume that the administrator has to complete a form with three input data: *Price*, *Description*, and *VAT*. The *Price* is required to be ≥\$0.01. This constraint leads to the definition of a valid partition—that formed by all real numbers satisfying such a constraint—and one invalid partition—that formed by all other real numbers. In the case that the price is mandatory input data, the NULL value represents another invalid partition. The *Description* is a string of valid characters; thus, any permutation P of x characters or ciphers is valid. For the sake of brevity, we have considered only the NULL value as an invalid partition, but infact any combination including at least one invalid character ($c \notin \{a, \ldots, z, 0, \ldots, 9\}$) could be considered as a value of another invalid partition. Finally, assuming that there exist three *VAT* rates (e.g., 4%, 10%, and 22%), the set of VAT values is the valid partition. Consequently, any other value is within the invalid partition. Again, the NULL value is another invalid partition if

Table 10.1 Example of equivalence partitioning analysis

Input data: *Price*	**Values**	**Tag**
Valid Partition	$\{x\} : x \geq 0.01\$$	VP1
Invalid Partition	$\{x\} : x < 0.01\$$	IP1
Invalid Partition	{NULL}	IP2
Input data: *Description*	**Values**	**Tag**
Valid Partition	$\{P(x)\} : x \in \{a, \ldots z, 0, \ldots, 9\}$	VP2
Invalid Partition	{NULL}	IP3
Input data: *VAT*	**Values**	**Tag**
Valid Partition	{4%, 10%, 22%}	VP3
Invalid Partition	$\{x\} : x \notin \{4\%, 10\%, 22\%\}$	IP4
Invalid Partition	{NULL}	IP5

VAT is mandatory input data. Table 10.1 shows the equivalence partitions derived from the previous specifications.

After this, it is possible to design a test suite with the aim of covering all the test conditions; i.e., from each valid or invalid partition, one test value is chosen. Once a test condition has been exercised with one value, the behavior of the system is assumed to be the same against all other values of the partition. Table 10.2 reports some test cases designed with the equivalence partitioning technique. *TC1* uses input values of only valid partitions and enables the tester to exercise the test conditions VP1, VP2, and VP3. Subsequently, a new test case is specified with the aim of covering at least one new test condition. In the table, *TC2* adds the condition IP4 to those already covered. An incremental approach is used (designing a new test case) as long as there are uncovered test conditions.

10.2.2 Boundary value analysis

Boundary value analysis integrates with the equivalence partitioning technique. The leading idea is that mistakes and bugs are more frequent at the partitions' boundary values.

It is common to adopt both techniques together. Indeed, on the one hand, the value at the boundary is also a value of the partition; therefore this value can be chosen to verify the behavior of the system using both testing techniques with one test. On the other, in any case in which the test should fail, it is not possible to assert that the system will fail for all the values of the partition but only that it may fail at the boundary; thus, another test with a different value taken from the partition is necessary.

Table 10.2 Example of equivalence partitioning test case design

Test Case: *TC1*	**Input values**	**Expected result**	**New covered conditions**
Price	$5.50	New product	VP1, VP2, VP3
Description	Home-made marmalade	Inserted	
VAT	4%	Correctly	
Test Case: *TC2*	**Input values**	**Expected result**	**New covered conditions**
Price	$5.50	Exception:	IP4
Description	Home-made marmalade	INVALID_VAT	
VAT	3%		
Test Case: *TC3*	**Input values**	**Expected values**	**New covered conditions**
Price	$5.50	Exception:	IP5
Description	"Home-made marmalade"	NULL_VAT	
VAT			
...	...	...	...
Test Case: *TCN*	**Input values**	**Expected values**	**New covered conditions**
...	...	...	...

Table 10.3 Example of boundary value analysis

Input data: ***Price***	**Values**	**Tag**
Valid Boundary	$x = \$0.01$	VB1
Invalid Boundary	$x = \$0$	IB1
Invalid Boundary	NULL	IB2
Input data: ***Description***	**Values**	**Tag**
Valid Boundary	$x \in \{a, \ldots z, 0, \ldots, 9\}$	VB2
Invalid Boundary	NULL	IB3
Input data: ***VAT***	**Values**	**Tag**
Valid Boundary	$x = 4\%$	VB3
Valid Boundary	$x = 10\%$	VB4
Valid Boundary	$x = 22\%$	VB5
Invalid Boundary	$x = 3.99\%$	IB4
Invalid Boundary	$x = 4.01\%$	IB5
Invalid Boundary	$x = 9.99\%$	IB6
Invalid Boundary	$x = 10.01\%$	IB7
Invalid Boundary	$x = 21.99\%$	IB8
Invalid Boundary	$x = 22.01\%$	IB9
Invalid Boundary	NULL	IB10

Concerning the example introduced in the previous section, Table 10.3 reports the boundaries of the partitions defined using the equivalence partitioning technique. The three partitions specified for the *Price* have one boundary each; we did not consider an upper limit for the price, nor a lower limit for the invalid price partition. Additionally, the NULL partition has a unique value, which is also the boundary. For the *Description*, we assumed as the boundary any string of one element and tested just one of these. Finally, the valid *VAT* partition is a set of finite elements, each one of which is also a boundary of the partition itself. To complete with invalid boundaries analysis, we had to approach each one of these values from both sides and added, as always, the NULL value.

Test cases are generally designed to exercise valid boundary values, and invalid input boundary values. Test cases may also be designed to test that invalid output boundary values cannot be induced. Table 10.4 shows the test cases designed to cover the test conditions identified in Table 10.3. *TC1* covers the conditions VB1, VB2, and VB5. In an integrated approach, of course, it would also cover conditions VP1, VP2, and VP3 of Table 10.1. All other test cases are designed and added one-by-time to exercise at least one new condition with respect to those covered by the previous test cases.

10.2.3 State transition testing

State transition testing is applied to finite-state components.

A finite-state component is characterized by a finite set of states. It has a finite set of inputs or events that can induce transitions between states. The behavior of the

Table 10.4 Example of boundary value test case design

Test Case: *TC1*	**Input values**	**Expected result**	**New covered conditions**
Price	$0.01	New product	VB1, VB2, VB5
Description	"H"	Inserted	
VAT	22%	Correctly	
Test Case: *TC2*	**Input values**	**Expected result**	**New covered conditions**
Price	$5.50	Exception:	IB4
Description	"Home-made marmalade"	INVALID_VAT	
VAT	3.99%		
Test Case: *TC3*	**Input values**	**Expected values**	**New covered conditions**
Price	$0	Exception:	IB1, VB2, VB3
Description	"H"	Incorrect price	
VAT	4%		
...	...	...	...
Test Case: *TCN*	**Input values**	**Expected values**	**New covered conditions**
...	...	...	...

component depends upon the current state and the input or event that occurs at that time.

Test cases are designed to exercise transitions between states. The number of valid transitions that a singular test case may exercise is unbounded; that is, the sequence of events generated by the test case can potentially be infinite. In addition to testing valid transitions, test cases should be designed to verify the robustness of the system by trying to force unspecified transitions.

The specification of a test case includes the starting state of the component, the input(s) to the component, the expected outputs from the component, and the expected final state.

A variety of coverage criteria applies to state-transition testing. As a minimum requirement, a tester should ensure that test cases visit every state and cover every transition. A potentially higher coverage criterion requires that at least one test covers each transition sequence of N, or less than N length, N being equal to 1, 2, or 3, or higher. This is also known as the "Chow's switch coverage" or "N-1 Switch Coverage."

Figure 10.1 shows a simplified state diagram for a room in a hotel. States are labeled with letters, and transitions are numbered.

Table 10.5 reports the cases of $N=1$ (0-Switch coverage) and $N=2$ (1-Switch coverage). Consequently, tests cases are designed as shown in Table 10.6. Note that the table reports all the test cases for 1-Switch Coverage (TC1–TC5) and an example of a test case for robustness checking (TC6).

10.2.4 Cause–effect graphing and decision table testing

Equivalence partitioning and boundary value analysis are very useful techniques especially when testing input field validation at the user interface. However, many verifications involve testing the business logic that stands beyond the user interface.

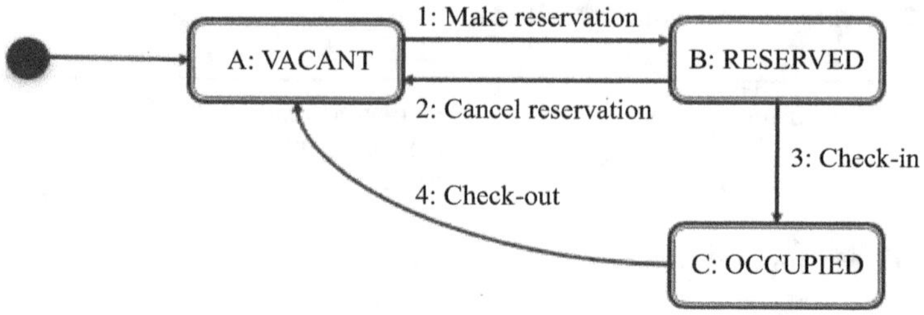

Figure 10.1 State diagram

Table 10.5 Example of N-Switch coverage analysis

Initial state	0-Switch sequences	1-Switch sequences
VACANT	A1	A1B2 A1B3
RESERVED	B2 B3	B2A1 B3C4
OCCUPIED	C4	C4A1

Table 10.6 Example of 1-Switch coverage test case design

Test Case: *TC1* A1B2	**Initial state** VACANT	**Sequence of input events** 1: Make reservation 2: Cancel reservation	**Expected result/new state** RESERVED VACANT
Test Case: *TC2* A1B3	**Sequence of input events** VACANT	**Input events** 1: Make reservation 3: Check-in	**Expected result/new state** RESERVED OCCUPIED
Test Case: *TC3* B2A1	**Sequence of input events** RESERVED	**Input events** 2: Cancel reservation 1: Make reservation	**Expected result/new state** VACANT RESERVED
Test Case: *TC4* B3C4	**Sequence of input events** RESERVED	**Input events** 3: Check-in 4: Check-out	**Expected result/new state** OCCUPIED VACANT
Test Case: *TC5* C4A1	**Sequence of input events** OCCUPIED	**Input events** 4: Check-out 1: Make reservation	**Expected result/new state** VACANT RESERVED
Test Case: *TC6* A2	**Sequence of input events** VACANT	**Input events** 2: Cancel reservation	**Expected result/new state** Exception: Forbidden operation
…	…	…	…
Test Case: *TCN* …	**Input values** …	**Expected values** …	**New covered conditions** …

Table 10.7 Example of cause-effect analysis

Causes								
C1	T	F	T	T	F	F	T	F
C2	T	T	F	T	F	T	F	F
C3	T	T	T	F	T	F	F	F
Effects								
E1	T	F	F	F	F	F	F	F

Cause–effect graphing and state transition testing are better tools for this kind of verification and validation.

Cause–effect graphing uses a model of the logical relationships between causes and effects. Each cause is expressed as a Boolean condition; i.e., a logic variable that is either true or false on an input, or combination of inputs. Each effect is specified as a Boolean expression that combines causes, in accordance with a business rule, using the logic operators AND, NAND, OR, NOR, and NOT.

The model may be described by a decision table representing the logical relationships between causes and effects.

Test cases are designed to exercise business rules, which define the relationships between the component's inputs and outputs.

To show an application of this technique, we will consider the case of cash withdrawal from an ATM. We will assume the following business rule for the withdrawal operation: "the withdrawal is approved when the client has inserted a valid PIN, the amount required is less than a daily limit, and the client's account balance is greater than the amount required." It is possible to distinguish three causes and one effect:

E1—The withdrawal is approved;
C1—The client has inserted a valid PIN;
C2—The amount required is less than a daily limit;
C3—The client's account balance is greater than the amount required.

Table 10.7 shows the decision table for the relationships among the previous causes and the effect. There is only one combination of Boolean values for the causes that enables the operation of withdrawal, the one for which all the causes are true. Such a truth table can be reduced as reported in Table 10.8.

Assuming that "123456" is a valid PIN and \$450 is the balance of the associated account, the test cases reported in Table 10.9 exercise all the test conditions (A, B, C, and D) defined previously; i.e., all useful combinations of the causes are emulated to verify the effect.

10.2.5 Syntax testing

Syntax testing uses a model of the syntax of the inputs to a component, which is typically defined by employing formally the Backus–Naur Form (BNF) notation. The syntax is specified through a set of rules.

Table 10.8 Example of cause–effect analysis (reduced truth table)

Causes				
C1	T	F	–	–
C2	T	–	F	–
C3	T	–	–	F
Effects				
E1	T	F	F	F
Tag	A	B	C	D

Table 10.9 Example of a cause–effect graphing test case design

Test Case: *TC1*	**Input values**	**Expected result**	**Test condition**
PIN	123456	E1	A
Required amount	$250		
Test Case: *TC2*	**Input values**	**Expected result**	**Test condition**
PIN	111111	NOT E1	B
Required amount	$250		
Test Case: *TC3*	**Input values**	**Expected result**	**Test condition**
PIN	123456	NOT E1	C
Required amount	$600		
Test Case: *TC4*	**Input values**	**Expected result**	**Test condition**
PIN	123456	NOT E1	D
Required amount	$400		

The leading idea is that the application inputs must be validated. Assuming that the data inputs conform to specific formats (e.g., textual formats, table schemas, file formats, etc.), test cases are designed to exercise both valid and invalid syntax.

For example, we consider the format of car number plates that in a certain country are coded by a couple of capital letters, representing the identification code of the regional area, followed by six digits (e.g., KC125344). Among the other formalisms, BNF provides the following:

"::=" means "is defined as";
"|" means "OR";
"K^n" means "*n* repetitions of *K*."

In Table 10.10, the syntax for the number plates is formally defined and some test cases for valid and invalid syntax are reported.

10.2.6 Combinatorial test techniques

The behavior of a software system may be affected by many factors, such as input values, preconditions, the internal state, and the execution environment.

Table 10.10 Example of a syntax testing rules and test case design

Data	**Syntax formally defined**
Capital_Letters	::= A | B | …| Z
Digits	::= 0 | 1 | 2 | 3 | 4 | 5 | 6 | 7 | 8 | 9
Regional_Code	::= Capital_Letters^2
Progressive_Code	::= Digits^6
Car_Number_Plates	::= Regional_Code Progressive_Code
Test cases	
Valid syntax	KC125344
Invalid syntax	1C125344
Invalid syntax	K1125344
Invalid syntax	KC1G5344
Invalid syntax	KC12534
Invalid syntax	K125344
…	

Testing techniques like equivalence partitioning and boundary-value analysis can be used to test the behavior of the system against possible values of individual factors. However, it has been demonstrated that most faults are caused by a combination of factors. On the other hand, exhaustive testing would require that the system is exercised with all possible combinations of factors. This is impractical for nontrivial systems. Indeed, if we consider a simple application that has eight input parameters, each of which being able to assume four values, the number of possible combinations is $8^4 = 4.096$.

Combinatorial testing relies on the consideration that not every factor contributes to every fault. On the contrary, often a failure is caused by interactions among a few factors. Consequently, instead of verifying all combinations, only a subset of these is considered. Combinatorial design can result in a drastic reduction in the number of combinations to cover, but assures a very effective fault detection.

By definition, an N-way interaction fault is a fault that is triggered by a certain combination of N input values. A simple fault is a 1-way fault, a pairwise fault is a two-way fault, a fault caused by the interaction of three different parameters is a three-way fault, etc. Since a majority of software faults consist of simple and pairwise faults, combinatorial testing techniques allow the detection of a greater number of faults.

Combinatorial Testing techniques include the following:

All combinations coverage—Every possible combination of values of the parameters must be covered;

Each choice coverage—Each parameter value must be covered in at least one test case;

Pairwise coverage—Given any couple of parameters, every combination of values of these two parameters must be covered in at least one test case;

Table 10.11 Each choice coverage

A	B	C
a1	b1	c1
a2	b2	c2
a1	b3	c2

Table 10.12 Pairwise coverage

A	B	C
a1	b1	c1
a1	b2	c1
a1	b3	c1
a1	–	c2
a2	b1	c2
a2	b2	c2
a2	b3	c2
a2	–	c1

Table 10.13 Base choice coverage

A	B	C
a1	b1	c1
a2	b1	c1
a1	b2	c1
a1	b3	c1
a1	b1	c2

***K*-wise coverage**—Given any set of *K* parameters, every combination of values of these parameters must be covered in at least one test case;

Base choice coverage—For each parameter, one value is selected as the base choice of the parameter. A base test is formed by using the base choice for each parameter;

Multiple base choices coverage—For each parameter, at least one value (but possibly more) is selected to be the base choice.

For example, if we consider three parameters $A = \{a1, a2\}$, $B = \{b1, b2, b3\}$, and $C = \{c1, c2\}$, all combinations coverage would require 12 tests. Differently, possible combinations for each choice coverage, pairwise coverage, and base choice coverage are reported in Tables 10.11–10.13. The test set chosen for each choice covers all possible values for the three parameters with at least one test. In the case of base choice, the values (a1, b1, c1) were chosen as base values. The test set is obtained by keeping always two of the three parameters equal to the base value and varying the third between nonbase values. Finally, a possible set of pairwise combinations is reported. Tools, such as VPTag [36] allows the determination of sets of all-pairs combinations automatically.

10.2.7 Scenario testing and use case testing

Use case testing uses the use case model of the system to derive test cases that exercise the whole system on a transaction by transaction basis from start to finish.

Use case view is part of UML modeling in which two main entities are involved: Actors and Use Cases.

Actors are either people or other artificial systems that solicit the system's functionalities and invoke a business service to the system.

A use case is a particular use, functionality, or feature of the system available to the actors. Each use case is described by means of sequences of interactions between the actor and the system, called scenarios. At least a main (successful) scenario is specified, which allows the achievement of the business goal associated with the use case. It may be useful also to specify some extensions of the main scenario, which describe interactions in any case in which the main goal cannot be achieved directly, but alternatives are available.

Use cases describe the process interaction flows between the system and its users against its most likely use. As a consequence, the test cases derived from use cases are particularly useful for finding defects in the real-world use of the system.

Each use case must specify all the preconditions that must be verified before. Use cases must also specify post conditions, which are observable results, and the final state of the system after the use case has been executed successfully.

Use cases serve as the foundation for developing test cases, mostly at the system and acceptance testing levels.

The reservation use case for hotel rooms example shown below describes both successful scenario and one unsuccessful scenario. The actor (or role) "Reservation Maker" wants to make a reservation in a hotel. The successful scenario describes the interaction that takes place when the actor achieves the business goal. The extension describes an alternative in any case in which no room is available for the check-in/check-out dates provided by the actor.

Use case description

Use case—Reservation
Actor—Reservation maker
Goal—To make a reservation in a hotel

Main scenario

1. The reservation maker asks for a reservation
2. The system provides a calendar for the selection of the check-in and check-out dates
3. The reservation maker chooses the check-in and check-out dates
4. The system checks for availability

5. The system provides details of all kinds of room available in the period indicated by the actor
6. The reservation maker chooses a room and confirms the reservation

Extension

5.a No room is available, but the system provides some availabilities for other similar dates

The use case testing technique requires that at least a test of the success scenario and a test for each extension is performed.

10.2.8 Random testing

Random Testing, also known as Monkey Testing, does not use any model to derive test cases. Indeed, all of the test input values are generated randomly, typically by an automatic tool.

This technique provides an easy and inexpensive way to estimate software reliability. No complex test case design activity is performed. On the contrary, several test cases are generated very rapidly, and an early measure of the level of quality of the system is obtained.

Random Testing may be useful to assure a uniform distribution of the test cases against either the input data partitions or the code areas, which otherwise, using the other (structured) testing techniques, could remain only lightly tested. If the domain is well structured, an automatic generation of random test cases may integrate with other testing techniques and allow an intensification of the testing activity, but it may produce many unrealistic or redundant test cases.

10.3 Structure-based testing technique

White-box testing techniques use the control flow graph as a model for the analysis. For the remainder of this section, as an example, we will consider the simple control flow diagram depicted in Figure 10.2, which implements the following requirements:

R1 : It prints the product of two numbers, if both are positive;
R2 : It prints the not negative number, if one of the two is negative;
R3 : It prints nothing, if both are negative.

10.3.1 Statement testing

The objective of the statement testing technique is to cover as many statements of the source code as possible. The paths (P1, P2, and P3) depicted in Figure 10.3 allow a 100% statement coverage. In particular, the three test cases reported in Table 10.14 enable a coverage of all the paths.

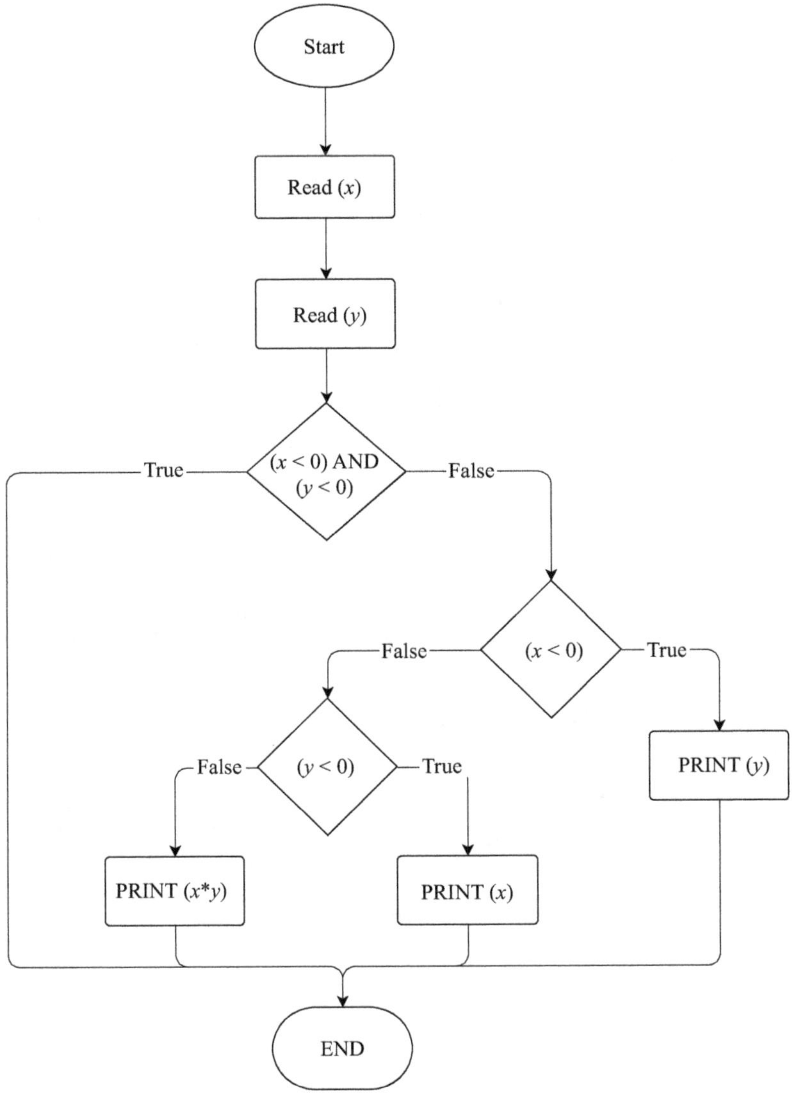

Figure 10.2 Flow chart example

10.3.2 Branch/decision testing

The branch/decision testing technique aims at covering as many decision outcomes as possible. A decision statement is an executable statement that has the capability of altering the sequence of instructions executed depending on the input data and the internal state of the system. Typically, decisions are found in selections and loops. A decision outcome is any possible path exiting from a decision statement.

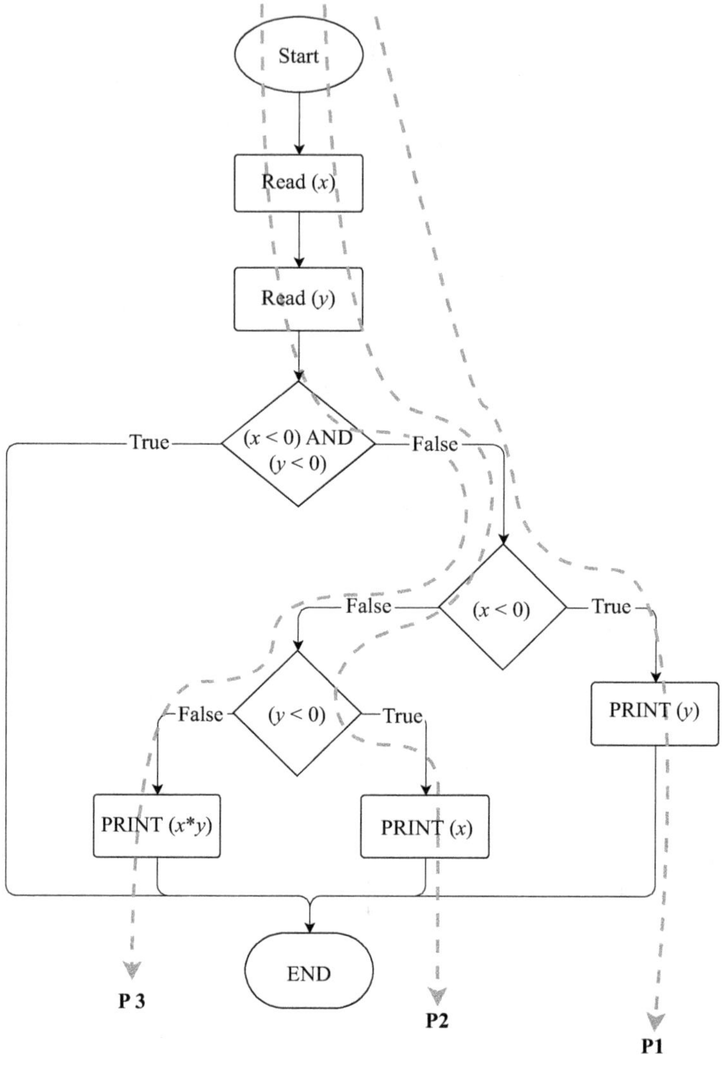

Figure 10.3 Statement testing

Test cases are designed to exercise decision outcomes. Considering the previous example, while all the decision outcomes are fully covered for both the second and the third decision, in the case of the first decision statement the "true" outcome is not covered. As a consequence, a further path is required to ensure a 100% branch/decision coverage: the one followed for any couple of negative inputs. The test case TC4 = {$x = -3, y = -2$; "NO VALUE PRINT"} may complete the test suite of Table 10.14 to achieve a full coverage.

Table 10.14 Example of a statement coverage test case design

Test Case: *TC1*	**Input data**	**Expected result**	**Path covered**
x	−1	1	P1
y	1		
Test Case: *TC2*	**Input values**	**Expected result**	**Path covered**
x	2	2	P2
y	−1		
Test Case: *TC3*	**Input values**	**Expected result**	**Path covered**
x	2	6	P3
y	3		

10.3.3 Condition testing

Condition Testing is a set of three specific techniques: Branch Condition Testing, Branch Condition Combination Testing, and Modified Condition Decision Testing. All such techniques focus on the Boolean decision expressions that may be the basis of a decision statement. As an example, we consider the statement "IF [A OR (B AND C)]"

Branch Condition Coverage requires that all the Boolean operands (A, B, and C) are evaluated individually, both TRUE and FALSE. Typically, a 100% Branch Condition Coverage can be obtained with two test cases, independently of the number of Boolean operands involved in the expression: the first with all true values and the second with all false values.

Branch Condition Combination Coverage, instead, requires that all possible combinations of the Boolean operands are evaluated.

Finally, Modified Condition Decision Coverage is a compromise between the two previous condition testing techniques. It requires fewer test cases than Branch Condition Combination Coverage and aims at showing that each Boolean operand can independently affect the outcome of the decision.

Tables 10.15–10.17 report the test conditions for all the Condition Testing techniques. The tests reported for the Modified Condition Decision Coverage technique are some of the possible suites obtained starting from the inclusion of a couple of test cases for each Boolean operand with the two possible values (true and false), then purging the set from repetitions (details can be found in [37]).

10.3.4 Data flow testing

Data flow testing uses the control flow model for the component and aims at testing control flow paths between definitions (Def) and uses (Use) of variables in the source code identified, as shown in Section 9.3.

Table 10.15 Branch condition combination testing

Case	A	B	C
1	TRUE	TRUE	TRUE
2	FALSE	FALSE	FALSE

Table 10.16 Modified condition decision testing

Case	A	B	C
1	FALSE	TRUE	TRUE
2	TRUE	FALSE	TRUE
3	FALSE	FALSE	TRUE
4	FALSE	TRUE	FALSE

Table 10.17 Branch condition testing

Case	A	B	C
1	TRUE	TRUE	TRUE
2	FALSE	TRUE	TRUE
3	TRUE	FALSE	TRUE
4	TRUE	TRUE	FALSE
5	FALSE	FALSE	TRUE
6	FALSE	TRUE	FALSE
7	TRUE	FALSE	FALSE
8	FALSE	FALSE	FALSE

10.4 Error-guessing testing technique

10.4.1 Error-guessing

Error guessing relies on the experience of the tester to foresee what kind of defects there may be in the system and in which area of the source code they may be located. It is not a structured approach. Instead, it is more an "art" of the tester/developer who guesses the presence of bugs. This technique is quite simple and inexpensive. The effectiveness depends on the level of experience of the tester.

As an example, an experienced tester may reveal defects by testing the system with very small or very high numeric input values, unusual input data, NULL values, or unusual combinations of input data.

Chapter 11
Formal verification

11.1 Introduction and background

Formal verification is another way to check several quality characteristics of a system. The aim is to explore all the possible states of the system to check for properties of interest such as safety (nothing bad will happen), liveness (something good will happen), and fairness (independent subsystems make progress), or common design flaws such as deadlock (the system should not reach a state in which no further action is possible), and livelock/starvation (when a subsystem is prevented from taking any action because of resource contention).

Deductive reasoning and model checking are two alternative approaches to formal verification. The former requires the specification of some desired properties as formulas, by means of which a proof system (i.e., a set of axioms and rules) allows the designer to prove that the system meets the expected behavior formally. This approach may be applied to verify infinite state systems. Generally, the procedure can be automated. However, no limit can be assumed on the amount of memory or time may be needed in order to verify the system (i.e., find a proof). Such a technique, indeed, is very time consuming and can only be performed by experts of logical reasoning.

The latter, **Model Checking**, was defined by Clarke and Emerson as "an automated technique that, given a finite-state model of a system and a logical property, systematically checks whether this property holds for (a given initial state in) that model" [38]. Model Checking verifies that some properties, expressed using a formal logic, are satisfied by a model of the system. The verification requires that all possible states of the model are explored. As a consequence, only finite-state transition systems can be verified. This approach is fully supported by automatic tools such as, for example, SPIN [39]. Every time the model violates a desired property, the tool produces a counterexample, which illustrates the behavior that caused the violation of the property. Model Checking allows the designer to analyze even partial specifications. Therefore, it is possible to start with the specification of critical functionalities and verify them as a first step.

The first attempt at formal verification was made by Pnueli who used linear temporal logic (LTL) to specify temporal properties that the system had to meet. LTL enables to reason about changes in the truth value of formulas over a linearly ordered temporal domain. A few years later, Clarke and Emerson proposed a Computation Tree

Logic (CTL), i.e. a Branching Time Logic, whose model of time is a tree; that is, there are different paths in the future, any one of which may be realized following a singular branch (a sequence of temporal events) of the tree. These two logics are somewhat complementary, as there are some temporal properties expressible in only one of the two, either CTL or LTL. More recently, other formal logics (e.g., Ambient Logic) have been proposed to model other kinds of properties such as, for example, spatial properties.

The main challenge in model checking is dealing with the state space explosion problem. This problem occurs in, for example, systems with many components that can interact with each other, systems with data structures that can assume many different values and systems that may require unbounded replication of processes. In such cases, the number of global states can grow uncontrollably and may lead to undecidability.

11.2 Formal specification

Formal methods are well-known techniques able to increase quality by eliminating defects at the requirements specification and early design stages. Indeed, on the one hand, they represent the most effective tool to specify requirements in an unambiguous way; that is, they improve fault prevention, which is one of the fault avoidance techniques. On the other, they provide means to build models of the system that can be verified automatically by model checkers.

11.2.1 Ambient calculus and ambient logic

Among formal methods, **Ambient Calculus** was proposed by Cardelli and Gordon to describe the movement of processes and devices [40]. In Ambient Calculus, all entities are described as processes. An important kind of process is an ambient, which is a bounded place where computation happens. An ambient delimits what is and what is not contained and executed within it. Each ambient has a name and can be moved as a whole, into and out of other ambients, by performing in and out operations. Ambient Calculus allows a description of complex phenomena in terms of the creation and destruction of nested ambients, and of the movement of processes into and out of these ambients.

In Ambient Calculus, $n[P]$ represents an ambient, n being the name of the ambient, and P being the process running inside it. The process "0" is the one that does nothing. Parallel execution is denoted by a binary operator "|" that is commutative and associative. "$!P$" denotes the unbounded replication of the process P within its ambient.

The following expression

$$n[P_1 | \ldots | P_j | m_1[\ldots] | \ldots | m_k[\ldots]]$$

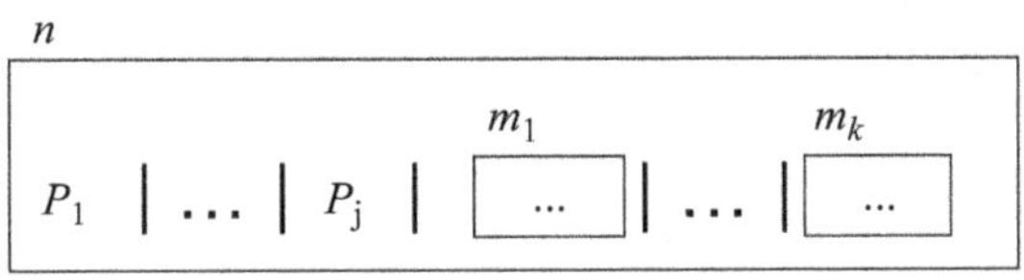

Figure 11.1 Graphical representation of ambients in Ambient Calculus

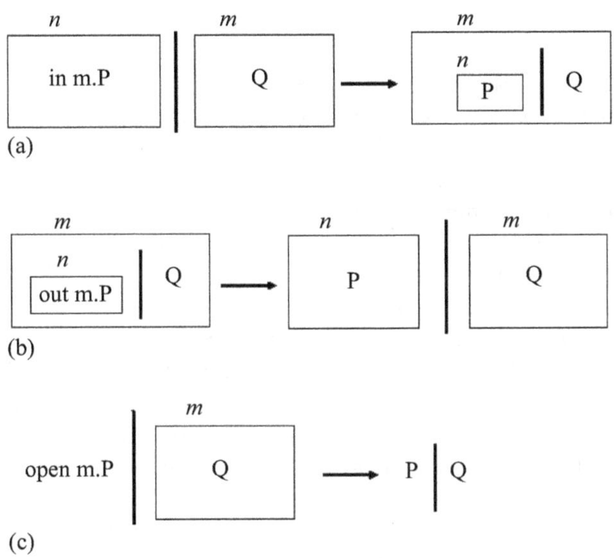

Figure 11.2 Ambient Calculus: capabilities. (a) in capability; (b) out capability; (c) open capability

describes an ambient with the name, n, that contains j processes with the names $P_1, \ldots, P_j$ and k ambients with the names $m_1, \ldots, m_k$. Figure 11.1 reports the equivalent graphical representation.

Some processes can execute an action that changes the state of the world around them. This behavior of processes is specified by using capabilities. The process *M*.P executes an action described by the capability *M*, and then continues as the process P. There are three kinds of capabilities, for entering (in), for exiting (out), and for opening up (open) an ambient, respectively. Figure 11.2 shows these three capabilities.

The process "in *m*.P" instructs the ambient surrounding "in *m*.P" to enter a sibling ambient named *m*. The reduction rule, which specifies the change in the state of the world is

$$n[\text{in } m.\text{P}] \mid m[\text{Q}] \rightarrow m[n[\text{P}] \mid \text{Q}] \tag{11.1}$$

The action "out m.P" instructs the ambient surrounding "out m.P" to exit its sibling ambient named m. The reduction rule is

$$m[n[\text{out } m.\text{P}] \mid \text{Q}] \rightarrow n[\text{P}] \mid m[\text{Q}] \tag{11.2}$$

The action "open m.P" provides a way of dissolving the boundary of an ambient named m located at the same level as open. The reduction rule is

$$\text{open } m.\text{P} \mid m[\text{Q}] \rightarrow \text{P} \mid \text{Q} \tag{11.3}$$

Finally, the restriction operator (Vn)P creates a new (unique) name n within a scope P. The new name can be used to name ambients and to operate on ambients by name.

Ambient Logic is a temporal (modal) logic that allows the expression of properties with respect to a specific location (ambient).

Logical formulas in Ambient Logic include (1) those defined for propositional logic, like negation ($\neg$), disjunction ($\wedge$), and conjunction ($\vee$); (2) universal ($\forall$) and existential ($\exists$) quantifications, as in first-order logics; (3) eventually ($\Diamond$) and always ($\Box$) temporal operators, as in temporal logics; and, in addition to classic logics, (4) spatial operators such as somewhere ($\triangledown$) and everywhere ($\triangle$).

The satisfaction relation $P \models A$ means that the process P satisfies the closed formula A (i.e., A contains no free variables). More specifically, the formula $P \models A@L$ means that the process P satisfies the closed formula A at the location L.

In the temporal modality, the relation $P \rightarrow P_0$ indicates that the ambient P can be reduced to P_0 by using exactly one reduction rule.

Similarly, in the spatial modality, the relation $P \downarrow P_0$ indicates that the ambient P contains P_0 within exactly one level of nesting.

Ambient Calculus and Ambient Logic enable the designer to easily describe a sequence of movements of users and devices (both entities are modeled as ambients) within a Smart Environment and its locations (again modeled as ambients) in terms of *in* and *out* operations. Additionally, it is convenient to model as ambients the data packages that move within a Body Sensor Network and toward the monitoring stations to be processed.

We have adopted Ambient Calculus and Ambient Logic to specify some of the requirements of an Intelligent Monitoring System for a Department of Nuclear Medicine. Within the department, indeed, patients are injected with a small amount of radioactive substance in order to perform specific examinations (e.g., a Blood Volume Study, Bone Scan, or Brain Scan). Once injected, patients emit radiation (gamma rays) and have to stay in a specific room waiting until the examination in question can be performed. The duration of the patient's waiting time depends on the kind of examination that has to be performed and the time that the radioactive substance takes to propagate—within the body—and to decay to the right level. Indeed, such examinations can be executed only if the radiation level is within a certain range. After the examination, the patients return to the waiting room until the level of radiation is below a specific threshold and, therefore, is innocuous.

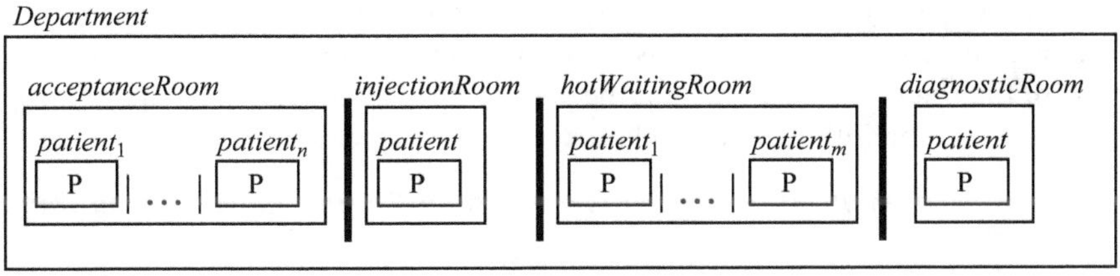

Figure 11.3 Structure of the Department of Nuclear Medicine

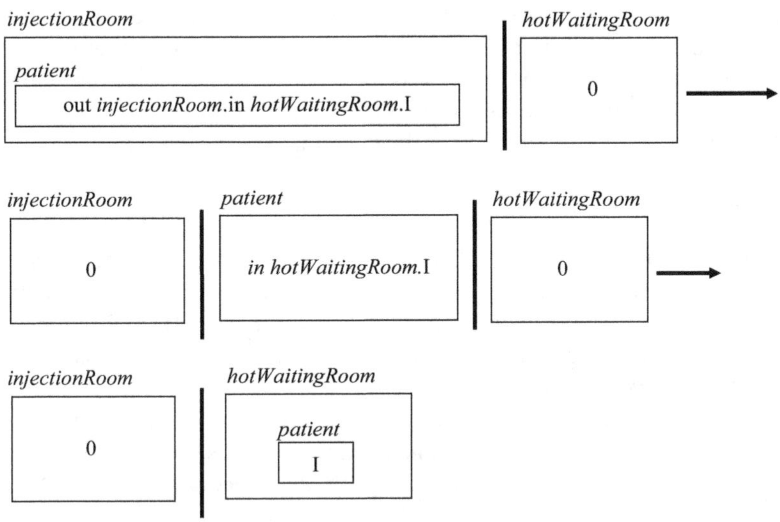

Figure 11.4 A section of the patient route within the department

Currently, these patients are called and accompanied by nurses within the department. The goal is to have an automatic system that leads the patients within the department and verifies the correctness of their movements.

The department (modeled as shown in Figure 11.3) consists of the following rooms:

- the *Acceptance Room*—this is the room where the patients are accepted within the department and wait for the injection;
- the *Injection Room*—this is the room where the patients (one by one) receive the injection;
- the *Hot Waiting Room*—this is the room where the patients wait for the examination after having been injected, remaining there until the radiation level reaches the correct range; and
- the *Diagnostic Room*—this is the room where the examinations are performed.

After the structure of the department has been modeled, it is possible to specify the precise route that the patient has to follow. For example, Figure 11.4 reports a section of the route: the movement of the patient from the Injection Room to the Hot Waiting Room. The following formula specifies the movement:

S_0: *injectionRoom*[*patient*[out *injectionRoom*.in *hotWaiting Room*.I]] | *hotWaiting Room*[0] →
S_1: *injectionRoom*[0] | *patient*[in *hotWaitingRoom*.I] | *hotWaitingRoom*[0] →
S_2: *injectionRoom*[0] | *hotWaitingRoom*[*patient*[I]]

At S_0, the patient has to move out the Injection Room and successively into the Hot Waiting Room. At S_1, she/he will have exited the Injection Room. At S_2, she/he will have reached the Hot Waiting Room. The full specification of the system includes the movements of other entities (e.g., messages sent to the patient's smartphone) and integrates some constraints such as the duration of the waiting time in the Hot Waiting Room before the examination and after the examination.

11.2.2 Linear temporal logic

LTL introduces temporal operators that allow the designer to specify and verify properties relating to how a system changes over time. Such a logic allows the specification of properties as logical formulas. Operators include (1) those defined for propositional logic, like negation ($\neg$), disjunction ($\wedge$), and conjunction ($\vee$); (2) those defined to quantify, like universal ($\forall$) and existential ($\exists$) quantifications, as in first-order logics; and (3) those defined to express temporal constraints, like true in the next moment ($\bigcirc$), sometimes true in the future ($\diamond$) and always true in the future ($\square$).

The satisfaction relation $P \models A$ means that the process P satisfies the closed formula A.

LTL may be adopted to specify properties such as safety, liveness, and fairness.

Figure 11.5, for example, reports the state diagram of a piece of software that handles the mutual exclusion for a critical section between two processes. The following are examples of some of the formulas that can be verified for the model:

1. $S \models \square \neg (C_1 \wedge C_2)$, which is a safety property meaning that it is always true that the two processes are not in the critical section at the same time;
2. $S \models \diamond\, C_1$, which is a liveness property meaning that there will be a moment, sooner or later, in the future, in which Process 1 is in the critical section;
3. $S \models \square\, (A_1 \Rightarrow \diamond\, C_1)$, which is another liveness property meaning that whenever Process 1 reaches the A_1 state, there will be a moment in which it is in the critical section; and
4. $S \models \square \diamond C_1$, which is a fairness property meaning that it is always true that there will be a moment, sooner or later, in the future, in which Process 1 is in the critical section.

Intuitively, we can verify the model. It is quite simple to demonstrate that the first property holds. Indeed, the state (C_1, C_2) does not exist. The third property also holds because all routes including any state for which Process 1 is in A_1 will necessarily visit

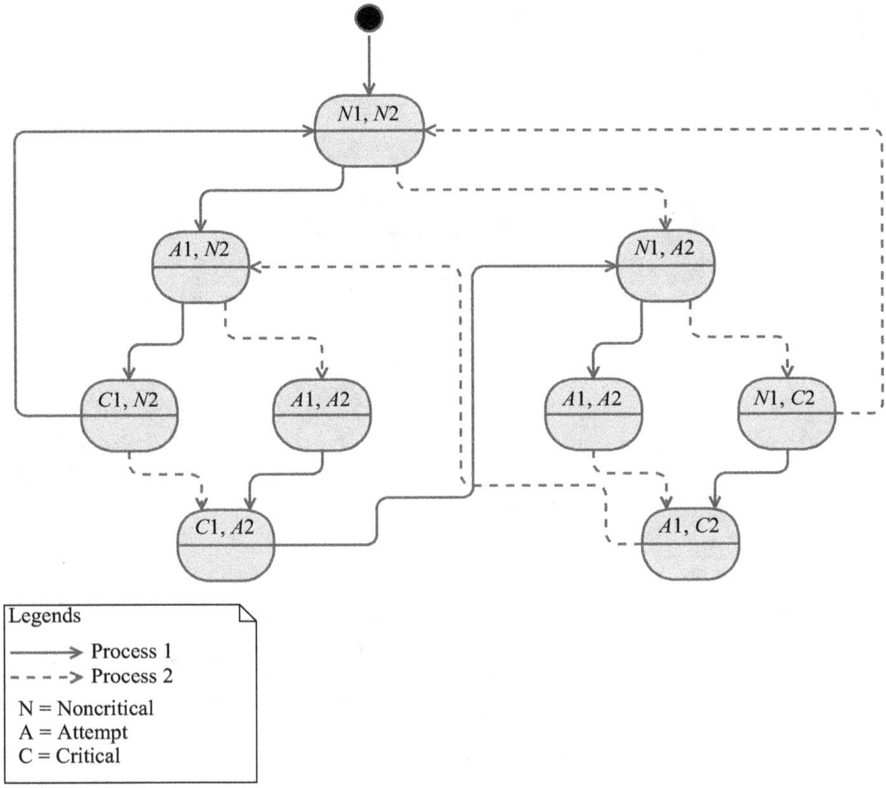

Figure 11.5 Mutual exclusion between two processes

(sooner or later) a state in which Process 1 is in C_1. Finally, the other two properties do not hold because the cycle $(N_1, N_2) \rightarrow (N_1, A_2) \rightarrow (N_1, C_2) \rightarrow (N_1, N_2)$ performed by Process 2 makes impossible for Process 1 to reach the critical section (C_1).

11.3 Model checking

The general structure of a Model Checking approach is shown in Figure 11.6, as presented in [41]. The requirements are formalized in a formal specification language such as LTL or CTL. The system model is processed by the Model Checker along with the property specification. The Model Checker will explore all the possible states of the system model, verifying if the properties hold for each of them. In any case in which a property is violated, at least one counterexample is generated to show the violation and assist in the debugging activity.

Although LTL is one of the most frequently used logics for model specification, model checking tools typically provide high-level languages for the specification of

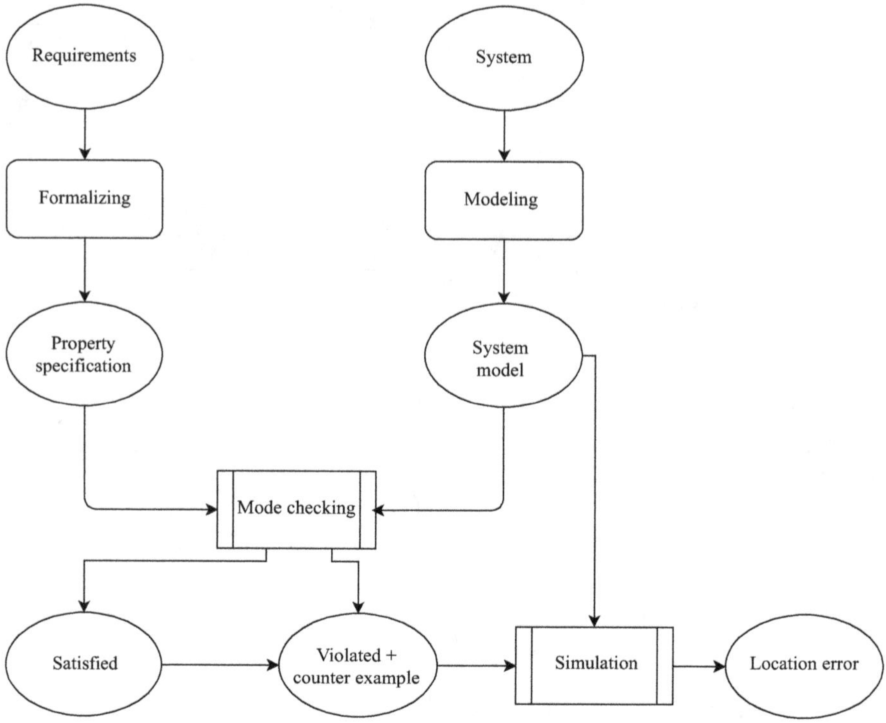

Figure 11.6 Schematic view of the model-checking approach

the system model and its properties. In the following section, we will present the major characteristics of Promela, which is the specification language for the SPIN model checker.

Promela is an imperative language specifically tailored for the purpose of writing high-level specification models of concurrent systems. Promela comprises three kinds of entities:

- Processes, which describe the structural components of the system;
- Data Objects, which are local or global variables; and
- Message Channels, which model the data exchange between processes, enabling both synchronous or asynchronous communication.

The semantics of the instructions is based on the concept of executability. Each statement is either executable or blocked depending on the system state. The selection among multiple executable statements is nondeterministic. However, it is possible to explicitly define blocks of uninterruptible sequences of instructions using the keyword atomic. The Promela language also provides flow control instructions, such as the if-else and the do repetition statements.

11.4 Static and dynamic (formal) verification

Static verification refers to proving the correctness of formally expressed specifications. Dynamic verification, also known as runtime verification, takes place during the execution and enables the system to check itself against correctness properties.

A relatively new direction of verification is runtime verification, defined in [42] as "the discipline of computer science that deals with the study, development, and application of those verification techniques that allow the checking of whether or not a run of a system under scrutiny satisfies or violates a given correctness property."

Runtime verification can be adopted to check correctness properties with respect to the system implementation. It involves the application of a lightweight formal verification during the execution of the system by checking traces of events generated from the system run against the correctness properties. When a property is violated, a recovery strategy is triggered. The technique scales well since just one model of computation is considered, rather than the entire state space as in model checking. Such a technique can be used both for testing prototypes before deployment and for monitoring the final system after deployment.

It is worth noting that static verification techniques, when they do not lead to problems like decidability or state space explosion, can prove the correctness of specifications. For this reason, they are considered to be stronger than runtime verification techniques. However, the correctness of specifications does not imply the correctness of implementation. This is particularly true for those applications whose behavior can strongly depend on the operating conditions. As an example, an Radio-frequency identification (RFID) reader can fail to detect a tagged user depending on, for example, (1) the orientation of the antenna, (2) the power of the antenna itself, (3) the position of the tag on body, or (4) the existence of an object between the tag and the reader at the time of reading. All these conditions are difficult to verify with static techniques. In contrast, runtime verification of the prototype system in the real domain of application can make possible the collection of traces that can be used to tune the system.

We expect runtime verification to become a major verification technique especially for systems in which the behavior of the system depends heavily on the environment and operational conditions. In other words, the behavior of highly dynamic systems, such as adaptive, self-organizing, self-healing, or pervasive systems, depends heavily on the environment and changes over time, making such behavior hard to predict and analyze prior to the execution. To ensure certain correctness properties, runtime verification can become part of the architecture of such dynamic systems. A noteworthy potential application for runtime verification is for the postmarket surveillance of medical devices.

11.5 Summary

A summary of the characteristics of the different verification techniques is reported in Table 11.1. As summarized in this table, both theorem proving and model checking

Table 11.1 Characteristics of the different verification techniques

Technique	**Characteristics** **Strengths**	**Weaknesses**
Theorem proving	Complete verification of the specification	State space explosion Possible undecidability
Model checking	Complete verification of the specification	State space explosion Possible undecidability
Testing specifications	No state space explosion	Partial verification of the specification
Runtime verification	Execution of recovery strategies Partial verification of the implementation	Need for additional components for the monitoring

can verify the entire specification, but may be subject to undecidability. Testing avoids any state space explosion but it enables only a partial verification. Finally, runtime verification allows only a partial verification and requires additional software components but can guarantee a verification of the running system (instead of its specification) and the execution of recovery strategies.

Part V

Techniques, methodologies, and engineering tasks for the development, configuration, and maintenance

Chapter 12
Prescriptive software development life cycles

12.1 Software as a product

The term software applies to computer programs, procedures, and documentation and data pertaining to the operation of a computer system.

It is a product of the human manufacture. However, it's different from all other kinds of product manufactured by humans because of a few peculiarities.

The software is immaterial. It is not consumed, but it deteriorates. If you bought the hardcover version of this book, sooner or later you will wear with use. On the contrary, you can use a software reader for the electronic version forever, without losing in quality. A software system, however, although it does not degrade in quality with the use, it does with the maintenance. Indeed, an empirical law of the software engineering asserts that all software systems deteriorate and loose in quality with the succession of maintenance operations, either corrective or evolutive, in time. This is because a maintenance operation may introduce new defects, even if it is a corrective one that fixes a particular bug. Moreover, the effectiveness of a maintenance operation degrees with the grow of the number of the maintenance operations that the system has already received; that is, maintenance-after-maintenance the software system moves far away from the original goal for which it had been conceived, designed, realized, and tested. This makes it difficult to keep the original level of quality and causes that the system is abandoned (sooner or later) in favor to a radically new major release or a completely new system.

The software is the result of a development process, which, however, is set up for producing only one sample of the "product." Other samples are obtained just copying and pasting the original. If we look at the first "development process" (i.e., assembly line) for the manufacturing industry, it was theorized by Frederick Taylor and implemented by Henry Ford who located workers along a belt to produce its famous T-model. Workers had synchronized tasks arranged in consecutive stages to perform. Any assembly line is set up to produce many samples of the same product. The development process for a particular software system, instead, is executed just one time. More important, the structure of the process may vary deeply type of software by type of software. For some classes of systems, a development process similar to an assembly line is sufficient. For other classes, development processes with a very different structure have been proposed and successfully adopted. On the other hand, software engineering concerns not only the systematic design and development of software products, but also the study and definition of new software development models.

Prescriptive models refer to a large and preliminary class of development process models that prescribe activities and task to be executed in a specific order. The activities are classified in structural activities and umbrella activities. The structural activities are mandatory and are synchronized with each other. Examples of structural activities are requirement analysis, design, coding, testing, etc.; that is, activities that we cannot skip to realize the software. The umbrella activities are not mandatory and may be executed in different moments of the development process. Examples of umbrella activities are risk management and quality control, which are not mandatory. Nonetheless, these may be extremely useful to achieve adequate levels of quality.

In the remaining part of the chapter, the most relevant software development strategies and prescriptive development process models are presented.

12.2 Software development strategies

The software development strategy determines the characteristics of the development process that will be chosen to conduct the project. The elements to be considered in order to choose the most appropriate development strategy are (1) the characteristics of the product; (2) the customer involvement in the product development; (3) the way the product is released to the client; and (4) the timing and costs of the development activities.

Regarding the characteristics of the final product, it must be considered whether they are immediately clear and consolidated rather than vague and unstable. The functional requirements may be ambiguous or unclear the beginning of the project. This will require a more intense communication activity with the customer. Moreover, the functional requirements may be unstable; that is, they are going to change during the development of the project. This will require that the communication with the customer is not confined to the preliminary phase, but it is spread along the development life cycle. Among the nonfunctional requirements of a product, it is of particular relevance the level of reliability and safety required. For example, in the case of safety-critical software, it is necessary to take into consideration such a particular characteristic, which will require a "certain amount" of validation and verification activities and task, thus affecting the life cycle, the project, and its costs.

The number of planned releases, the content each release and the modality of releasing, affect the development strategy. In fact, the development strategy for a product to be made available with a single issue will be different from that in which the functionality will be made available in multiple successive and independent releases and will still be different in the case each release provide new functionality and also improve those provided with the previous releases.

The degree of involvement of the client and other stakeholders in the project determines the manner some project tasks are managed. Revisions and approvals, participation in decisions, involvement in specific technical activities, and regular discussion of the progress of the project are just a few examples of tasks that are affected by the presence or not of the customer in the development team. The customer involvement helps to improve the interpretation of the requirements and allows the validation of the solutions but increases the project costs. On the contrary, a lack of involvement of the client reduces the project costs but requires that the requirements are very clear and stable already at the beginning of the project.

The initial available budget and timeslot and their level of flexibility are other elements that affect the development strategy. It is necessary to understand if costs and timetable have been planned considering the general high-level characteristics of the product, rather than stable and clear low-level requirements. Moreover, it must be considered whether or not the budget and timetable are fixed or can be re-modulated on-the-fly to cover possible changes during the realization. In the case costs are fixed, for example, any request for change requires negotiating in a "time/content-boxing" way the consequences; that is, each new required component can only be developed if replaces another of equal cost and time of realization.

The main software development strategies are:

One shot—The entire software product is realized in a single cycle. The release is unique and covers all the features and functionality of the system. The customer and all users typically participate only in the early stages of the project. This strategy can be adopted as long as all the requirements are clear, complete, and well stable since the beginning of the project. Furthermore, the set of system's requirements should be not particularly large.

Multishot—In this case, several development cycles are executed in succession. Each cycle focuses on a subset of the characteristics of the final product, but it can change what has been developed in previous cycles. This strategy is better than the One shot strategy when only a part of the product's features are very clear and stable. Indeed, in such a case the first cycles might focus on the clearest and more stable requirements giving the possibility, in the meantime, to go into details with other requirements.

Prototyping—Prototypes help to clarify the requirements through comprehensive interfaces of the system, or to explore the complexity and feasibility of critical functions, for example, for stringent performance or security requirements. The prototype does not have to evolve and become the system since it is not designed to maximize the product quality.

Phased incremental—This strategy defines several stages that are performed in different cycles. Each cycle produces a self-consistent increase of the product and allows the development team to handle large sets of system's requirements.

Evolutionary—It relies on subsequent releases of the product features. Once a release is issued, the implemented functionality may be refined in subsequent releases also as a consequence of the feedback gathered from the system running. This approach is to be preferred when the requirements are not stable or when the product needs continuous updates over time.

The development strategy affects the choice of the life-cycle model.

12.3 Waterfall models

12.3.1 The waterfall

The waterfall model is very popular for its simplicity. Historically, the early software development processes defined in software engineering were of this type. It is very similar to a traditional assembly line. In fact, the waterfall model defines both a sequence of phases (structural activities) and subactivities (tasks) and the results (deliverables) of each phase.

Figure 12.1(a) shows a specific development process of the class of the waterfall models. It must be noted that the sequence of phases can vary from process to process, as well as the tasks may become structural activities in other processes depending on the importance that these can have with regard to the specific product to realize. As an example, concerning the process shown in the figure, testing is a simple task. This, however, could be a structural activity in another process depending on the characteristics of the product to realize. Of course, the greater the emphasis is, the greater the effort required is.

In order to reduce the rigidity, a modified version of the model has been proposed (Figure 12.1(b)). This introduces new tasks for Verification&Validation at the end of each phase. Given a certain phase, it enables to return at the previous phase when a modification of the results obtained previously is necessary.

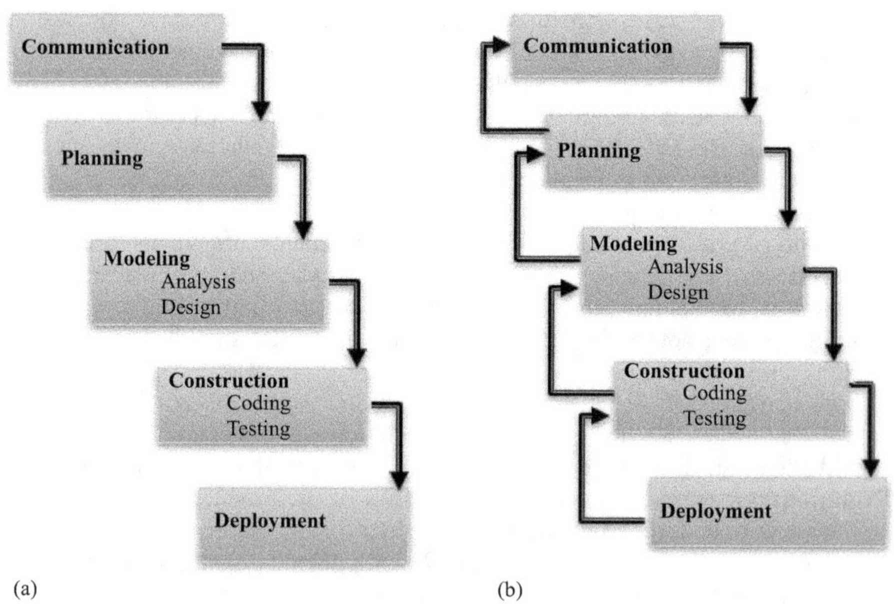

Figure 12.1 Waterfall model: (a) a waterfall process and (b) a modified waterfall process

The model provides a single release at the end of the development cycle. It relies on the waterfall development strategy. Consequently, it is useful with respect to projects for which the system's requirements are well understood, complete, and stable already at the beginning.

12.3.2 The V-model

The V-model is a software development model that can be considered a specialization of the waterfall. It is also applicable to the development of hardware.

The phases of the model are depicted in a V-shape. They are still consecutive as for the waterfall model, but during the first activities test cases are designed to be executed in the successive phases. In particular, the descending branch (left) of the V is dedicated to the definition of the system. It comprises activities for the conception, design, and realization. The ascending branch (right) of the V is dedicated to Verification&Validation. The V-model illustrates the relationships between each phase of the development branch (the descending one) and the testing phase associated with it (the ascending branch). The horizontal axis represents the time of the project, from left to right. The vertical axis represents the level of abstraction, with the lowest level of abstraction at the bottom.

The process depicted in Figure 12.2 comprises four definition phases and four testing and integration phases:

Requirements analysis—In this phase, the system requirements are gathered. It deals with determining the intended use. At this stage, some acceptance test cases are designed to establish a minimal level of functionality that the system must support to be considered complete.

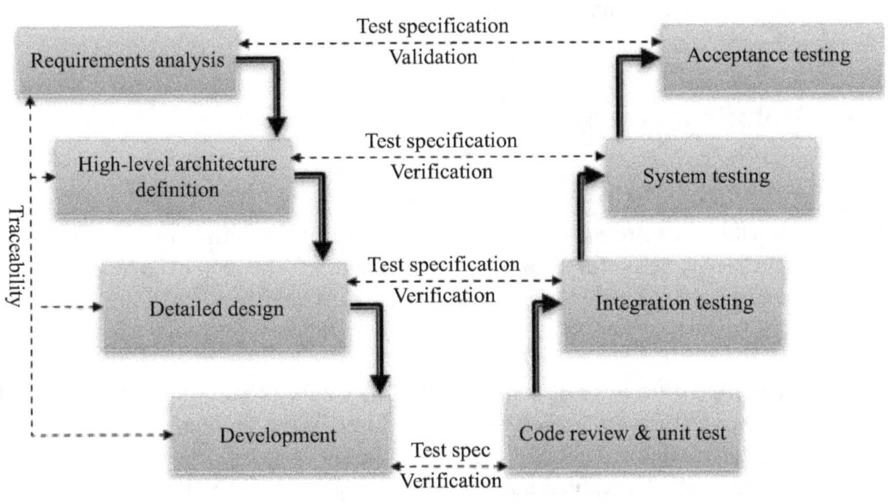

Figure 12.2 A V-model process

High-level architecture definition—In this phase, the high-level architecture is designed. It concerns the general organization of the system, the identification of subsystems and components, the dependencies between such components, the identification of use cases and actors, etc. In this phase, system test cases are designed.

Detailed design—This phase realizes detailed models of the system and its components. The system is divided into very small units, each of them is specified so that the programmer can immediately begin the coding. At this layer, the model may include a detailed functional description of the units, the database schema, all interface details, etc. In this phase, the unit test cases are defined.

Development—This phase concerns the coding of the software components.

Code review and unit test—In this phase, several verifications may be performed to ensure a certain level of quality of the singular software unit in isolation. The code may be reviewed through static analysis techniques. It is normally tested with white-box techniques.

Integration testing—After having tested the software units in isolation, during this phase cooperations between them are verified. Typically, an incremental integration approach is adopted; that is, units are integrated one-by-time. The testing activity is usually driven by black-box techniques.

System testing—This phase starts with the integration of the last software unit. The system testing is still an integration testing performed on a complete system and has the purpose of assessing the compliance of the system with the specified requirements. It is the final verification stage. The system testing is still driven by black-box techniques.

Acceptance testing—This phase concerns the validation of the requirements. It aims at demonstrating that the system meets the intended use and satisfies the customer. The acceptance testing allows the customer to determine whether or not to accept the system.

12.4 Evolutionary models

12.4.1 Prototype models

In any case in which requirements are unclear, or some functionality is very critical, the Prototype model may be useful to gather an early feedback and refine the requirements for the system. Figure 12.3 shows a prototype model, which consists of the following phases:

1. *Requirements gathering and analysis*. At this stage, requirements are gathered and analyzed as they are known at the beginning of the project;
2. *Quick design*. A quick design of the system based on the requirements analyzed so far is provided. The model concerns only major characteristics of the system;
3. *Building prototype*. The first prototype is realized at the first cycle. During the next cycles, such a prototype will be refined;

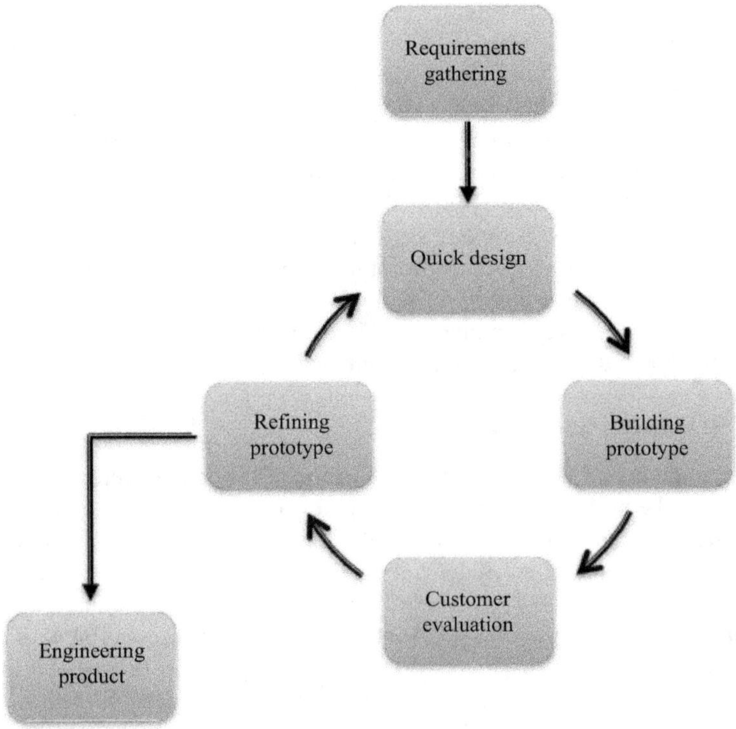

Figure 12.3 The spiral model

4. *Customer evaluation*. The prototype is presented to the user for her/his validation. This activity identifies the prototypes' strengths and weaknesses. The collected feedback is reported to the development team;
5. *Refining prototype*. The development team refines the prototype according to the user's feedback and new requirements collected. The short cycle *Quick design* → *Building prototype* → *Customer evaluation* → *Refining prototype* continues until the prototype does not meet the intended use and the customer is not satisfied by the final prototype;
6. *Engineer product*. When the prototyping cycle ends with a satisfactory prototype, the final system is realized starting from the last set of requirements that the final prototype has contributed to clarify.

There exist four types of prototypes:

Throwaway/rapid prototyping—This type of prototype is realized in a very short time and is discarded as soon as the requirements are completely understood. The prototype does not evolve into the system. In fact, it is not conceived with the objective of maximizing the customer satisfaction nor the quality of the final system.

Evolutionary prototyping—Evolutionary prototyping, which is also called Breadboard prototyping, is a minimal functionality based on actual requirements. It is often realized from the most critical requirements. It is not discarded, instead, it will be included in the final system, which is built adding new (noncritical) requirements to the prototype.

Incremental prototyping—This strategy allows the development team to realize multiple functional prototypes of the various sub-systems, which are finally integrated to build the complete system.

Extreme prototyping—Extreme prototyping consists of two major sequential phases. First, a basic prototype with all the interfaces is realized. Second, a prototype service layer is realized and integrated. The complete set of interfaces helps to clarify definitively the requirements for the system.

12.4.2 The incremental model

The incremental model is based on the realization of different development cycles in order to build the system incrementally. Each cycle, indeed, focuses on a subset of self-consistent features and allows the development team to create some functionality that are released to the client independently from those of the other development cycles.

A very popular (traditional) incremental development process is the Rapid Application Development (RAD). The RAD (Figure 12.4) provides unique stages for the initial communication and planning. Subsequently, several development cycles are implemented to realize the so-called "increments." Finally, the various increments are integrated into one system during an integration and deployed to the client during a subsequent phase. The RAD allows the reduction of the time of implementation of the system as long as different development teams may act in parallel in the different cycles defined to obtain the increments. It is particularly useful in large projects that need short time for the realization.

12.4.3 The spiral model

The spiral model is an evolutionary model that starts from a sub-set of the requirements and cycle after cycle, from the one hand, it extends the set of requirements treated in the cycle, and from the other, it allows the development team to refine all the previously developed requirements.

It is particularly useful in any case in which the requirements are unclear or unstable. In these situations, this model allows the development team to start treating a minimal set of requirements (those clearest and more stable) and to obtain an early and rapid prototype. The prototype system is used soon to gather the user's feedback, which is then used to refine the requirements in order to make the prototype evolve to the final product iteration after iteration.

The process shown in Figure 12.5 is useful for the medical software community. Indeed, it places much emphasis on risk management. Each of the four quadrants represents a structural activity of the process: planning, risk analysis, engineering, and evaluation.

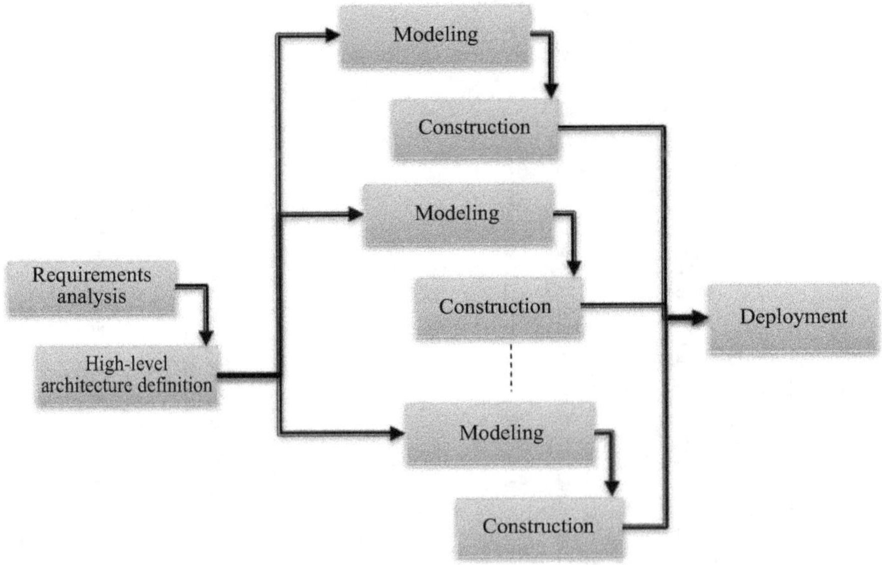

Figure 12.4 The Rapid Application Development process

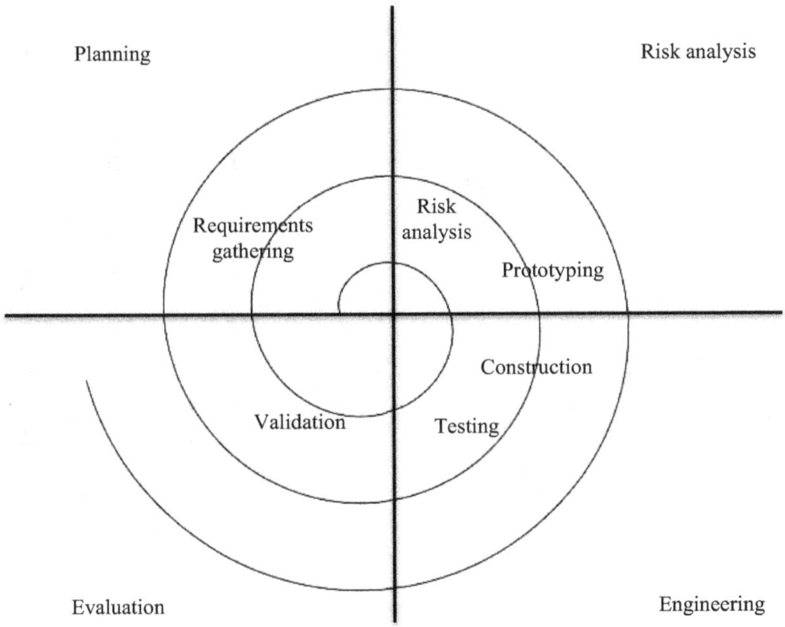

Figure 12.5 The spiral model

Table 12.1 Main characteristics of the described prescriptive models

Waterfall	**Advantages**	**Disadvantages**
	Simple Well established	Not effective for complex systems Requirements must be clear and stable already at the beginning Very rigid Communication with the client only at the beginning Delivery only at the end
V-Model	Advantages Simple Well established A lot of V&V	Disadvantages Not effective for complex systems Requirements must be clear and stable already at the beginning Very rigid Communication with the client only at the beginning Delivery only at the end
Prototype	Advantages Early user feedback Critical functionality are verified as first Unclear requirements are clarified Customers are involved in the development Defects may be detected early Critical functionalities are verified as first	Disadvantages The focus in not on quality The prototype often is slightly documented Risk of requirement analysis biased by the prototype's requirements
RAD	Advantages Shorten development time Good for large sets of requirements Frequent deliveries and communications with the client	Disadvantages Needs for independent development teams Requirements must be substantially clear and fixed
Spiral	Advantages A lot of risk analysis Manages well both unclear and unstable requirements Frequent interactions with the client	Disadvantages It does not work well for small-sized projects

This process is quite simple, but the number of spirals may grow as the complexity of the project grows.

12.5 Choosing the best software development model

Typically, there is not a unique development model that can be chosen to realize a particular software system. However, the properties of one development model may meet better than others the characteristics of the system to realize. The following principles may be taken into account to choose the development model:

1. Choose the waterfall only if the system is simple and all the requirements are clear and stable.
2. If the time for developing the system is stringent and it is a major project constraint, prefer the RAD.
3. If the system's characteristics are unclear at the beginning, choose the prototype model, which enables to clarify the requirements, or a spiral model that allows the development team to refine the requirements over the time.
4. If the system is critical for the user's safety, choose the V-model, which enables to build the confidence in the system by means of recurrent verification and validation activities, or the Spiral model shown in Figure 12.5 that includes structural risk management activities.

Some more details about the diverse software development models are reported in Table 12.1.

An interesting study that represents one motivation (not certainly the only) for the presence of the next chapter in this book is the one conducted in [43]. It concerns, as reported in Table 12.2, the resolution of over 10,000 projects from 2011 to 2015 included in the CHAOS dataset. Such projects have been segmented by the waterfall model and the agile process.

Table 12.2 Project resolution by Agile versus waterfall

Size	Method	Successful (%)	Challenged (%)	Failed (%)
All size	Agile	39	52	9
Projects	Waterfall	11	60	29
Large size	Agile	18	59	23
Projects	Waterfall	3	55	42
Medium size	Agile	27	62	11
Projects	Waterfall	7	68	25
Small size	Agile	58	38	4
Projects	Waterfall	44	45	11

Chapter 13
Agile software development life cycles

13.1 The Agile Manifesto

The Agile Software Development Alliance emerged during a 3-day work meeting held on February 2001 in Utah, where 17 methodologists finally summarized their ideas about new ways of developing software systems and managing related projects in the Agile Manifesto [44].

At the beginning, the Agile Manifesto states the purpose of the movement: "We are uncovering better ways of developing software by doing it and helping others do it. We value:

> **Individuals and interactions** over processes and tools.
> **Working software** over comprehensive documentation.
> **Customer collaboration** over contract negotiation.
> **Responding to change** over following a plan."

The Agile Alliance is not a sect of anarchists who intend to subvert "the established order" of traditional prescriptive methods. Rather, it is a group of methodologists, intensely engaged in software development, which aims to bring out "higher value" through agile methods.

The first statement, individuals and interactions over processes and tools, does not mean diminishing the importance of the latter. Rather, it recognizes that the individual is the brain that plans and controls a process and the engine that runs it. Hence individuals have experience, ability and genius that should not be limited by the rigidity of the process. This will empower all participants in the development of a project, who are involved as "active parts" rather than mere executors of actions required by a process.

The documentation is, of course, an important asset in the support of a software system especially in terms of facilitating any corrective and evolutive maintenance operation, as well as of providing evidence of the quality of the product itself. However, the focus must be reported on the "working software." The development team members must not be obsessed by the need to produce comprehensive documentation. They should be left free to choose which documents to produce on the grounds that they are deemed necessary or useful.

The negotiation of contracts is still an action evidently necessary, but the interaction with the client must not be limited to this activity and to the initial gathering of requirements. The interactions with the customer must be continuous. Indeed, the customer should possibly be a member of the development team. Constant collaboration with the customer on the one hand allows you to gradually gather requirements and regularly check the system increments, but also helps to build mutual trust and to understand and address the difficulties that both sides can find over time.

Finally, planning continues to be a crucial activity in the realization of a software project. However, rather than making a detailed long-term plan, the team should plan long-term objectives and more detailed plans only for a shorter period. In any case, it should be assumed that flexibility and adaptability to change are essential characteristics in any planning process.

The Agile Manifesto also introduces 12 principles:

- **Our highest priority is to satisfy the customer through early and continuous delivery of valuable software**
  The focus must concentrate on customer satisfaction which is not obtained by signing detailed contracts or producing exhaustive documentation. Rather, customer satisfaction grows with the early and continuous delivery of new pieces of the system that support her/him in her/his business function.
- **Welcome changing requirements, even late in development. Agile processes harness change for the customer's competitive advantage**
  Changes must be taken into account due to both continuous feedback coming from the customer and the evolution of the market and improvements in technology. Agile methodologies aim at facilitating change in a more effective way, instead of attempting to prevent it.
- **Deliver working software frequently, from a couple of weeks to a couple of months, with a preference for the shorter timescale**
  The Agile Alliance is concentrated on reducing the delivery cycle time. It is worth noticing that a "delivery" is not simply a new release. A delivery should provide new business value for the customer, rather than just a set of improved functionality or a version of the software system fixed against some defects found previously.
- **Business people and developers work together daily throughout the project**
  A radical change in the requirements management process is encouraged in accordance with Agile methods. As a matter of fact, Agile designers do not expect a detailed set of requirements at the very beginning of the project. A high-level view of the requirements is initially envisioned. Such a preliminary set of requirements is subject to frequent change thanks to continuous interactions between the business people and the developers. The frequency of interactions may be surprising for non-Agile experts: it tends to happen daily.
- **Build projects around motivated individuals, give them the environment and support they need and trust them to get the job done**
  A motivated team is a critical success factor for any project, more important than the process, the strategy and the policy adopted for the development. People should be provided with all the tools and resources necessary to undertake their

work successfully, do successfully feel the confidence of the project manager in their actions and decisions.

- **The most efficient and effective method of conveying information with and within a development team is face-to-face conversation**
While prescriptive processes have stressed the requirement for formal documentation before starting a project and during its development, in the Agile community, it is recognized that the most effective and rapid way to convey information is by means of direct interactions. Of course, the assertion is not that of providing no documentation, rather than it is necessary to find the right blend between the information transferred via documentation and the information transferred via oral conversations. There is some information that may be difficult to convey by means of a written document. On the other hand, some concepts may need a formal (written) specification.
- **Working software is the primary measure of progress**
The paradigm of Continuous Integration requires that members of a development team integrate their work very frequently, as often as several times daily. Integration generally includes automatic testing and verification, so providing a measure of progress and early feedback.
- **Agile processes promote sustainable development. The sponsors, developers, and users should be able to maintain a constant pace indefinitely**
As stated by Martin Fowler and Jim Highsmith [45] , "Our industry is characterized by long nights and weekends, during which people try to undo the errors of unresponsive planning. Ironically, these long hours don't actually lead to greater productivity ... Sustainable development means finding a working pace (40 or so hours a week) that the team can sustain over time and remain healthy."
- **Continuous attention to technical excellence and good design enhances agility**
System design is distributed among several iterations. Agile processes still require design activities that are short and alternated frequently with code construction, verification, and refactoring. Hence, design is another continuous activity performed throughout the project. The project design is enhanced continually, iteration by iteration, throughout the project.
- **Simplicity—the art of maximizing the amount of work not done—is essential**
Processes should be kept simple so allowing members of the development team to exploit fully their genius and creativity. Rules in the process should be established only to regulate very common needs or critical situations.
- **The best architectures, requirements, and designs emerge from self-organizing teams**
Self-organization of the development team and continuous interaction and refinement/refactoring of increments lead to the best architectures and design patterns.
- **At regular intervals, the team reflects on how to become more effective, then tunes and adjusts its behavior accordingly**
As no process, either agile or prescriptive, represents the "universal solution," it is essential that the team of an agile development process periodically meets and discusses issues, such as "What has gone ok? What has not gone ok? How can we

improve our interactions? How can we improve our process? How can we improve the quality of our increments?" etc. By posing questions and reflecting on answers the team matures its experience and refines its operation. The final question is "How mature is our team in applying successfully the process and how can it be improved?" The aim is a sort of capability maturity model self-assessment.

13.2 Scrum

Scrum is an agile project management framework introduced in 1999 by Ken Schwaber [46]. Being a management framework, Scrum aims at increasing the productivity and creativity of the development team. It is not a development process nor a programing technique. Scrum leaves decisions about the programing techniques to the development team. Moreover, it does not prescribe any development activity. On the contrary, it is a framework within which one can employ various processes.

In the Scrum vision, three pillars uphold a radically new way, highly adaptive, to manage project development: transparency, inspection, and adaptation. Transparency concerns the visibility of significant aspects of the process, like the progress of the project development, to the entire development team and the client. Inspection is related to the determination to provide Scrum users with the means to inspect frequently the artifacts and the progress of the work in order to undertake control measures in any case of deviation with respect from the planned objectives. In such a case, adaptation requires the carrying out of countermeasures as soon as possible in order to mitigate the effect of any such deviations.

The Scrum big picture shown in Figure 13.1 summarizes most of the concepts and instruments. The **Product Backlog** is the list of all the known requirements to realize and results to obtain in order to complete the project successfully. It includes functional and nonfunctional requirements, as well as work items such as the development environments for the set-up and the test suite for the designing and execution. Such features are prioritized. The preliminary Product Backlog represents a long-term planning, which however is dynamically updated. A short-term planning is represented by the **Sprint Backlog**. The Sprint Backlog is obtained by selecting a certain number of features from the Product Backlog. The development team members and the Product Owner choose the features to realize at the beginning of each **Sprint**, which is a short, fixed-length iteration. A Sprint lasts typically from 2 to 4 weeks and is expected to produce a potentially shippable product with an appreciable value for the customer. For every Sprint, the concept of "Time-box" is applied: no disturbance is allowed to modify the duration or goal of the Sprint. In a case in which the team realizes that the Sprint Backlog cannot be implemented entirely during the Sprint, some requirements are removed and sent back to the Product Backlog.

13.2.1 Roles

The Scrum framework defines three roles: a Scrum Master, a Product Owner, and the Team. The **Scrum Master** is mainly in charge of ensuring that the process and

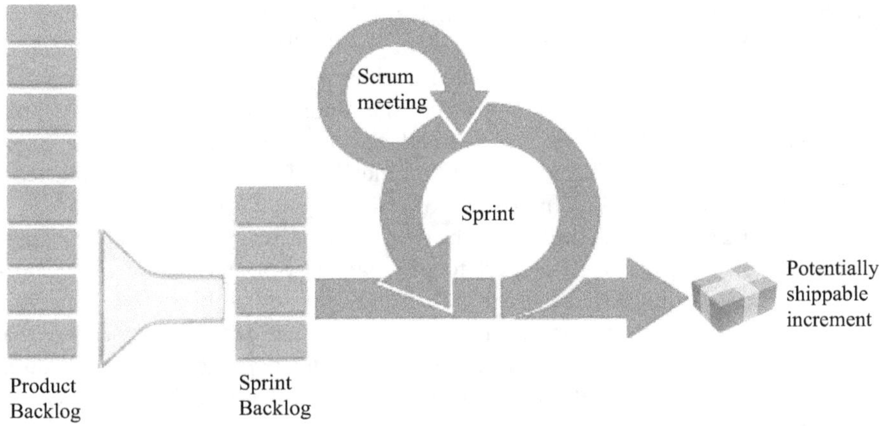

Figure 13.1 The Scrum framework

the methodology run correctly. She/he is also responsible for removing any obstacles that may impact on the productivity and for organizing and moderating meetings.

The **Product Owner** is responsible for maximizing the value of the product and the work of the Team. She/he is the sole person responsible for managing the Product Backlog by (1) listing and prioritizing the Product Backlog items; (2) ensuring that the Product Backlog is visible, transparent, and clear to all; and (3) ensuring that the Team understands the items in the Product Backlog to the necessary level.

The (Development) **Team** is small-sized and "cross-functional"; i.e., it is typically composed of from five to nine people with different skills such as analysts, designers, developers, and testers. The underlying idea is that a limited number of people may maximize the interaction, keeping the coordination simple and optimizing flexibility, creativity, and productivity. Of course, large projects may require the involvement of a greater number of people. In such a case, the Scrum scales up by means of an approach called the Scrum of Scrum. A hierarchy of teams is defined: at the bottom layer each team elects a representative member for a team at the upper layer, which in turn chooses a representative to be involved in a team at the upper layer, and so forth. A Sprint Backlog for a team at a certain layer is spread between the teams connected at the lower layer.

13.2.2 Events

The Scrum framework provides for the scheduling of a series of events to ensure regularity in the delivery of the increments and in evaluating the results. These events are specifically designed to enable the critical stages of transparency and inspection.

Before starting with a Sprint, the work to be performed is planned at the **Sprint Planning**. The Scrum Team (i.e., the Development Team plus the Scrum Master and Product Owner) conceives collaboratively the plan. Such an event is time-boxed to a maximum of 8 h for a 1-month Sprint, with a shorter time period possible especially in the case of shorter Sprints. The Product Owner presents the objective desired in the current Sprint, while the Development Team drafts the functionality to develop during the Sprint. The main input to this meeting is the Product Backlog, along with information on the past performances of the Development Team increments previously realized. The Development Team chooses all the items to be inserted in the Sprint Backlog based on the confidence it has in with a shorter time period possible. Next, the Scrum Team defines the Sprint Goal and the Development Team establishes a "Done" criteria for the product Increment to be developed during the Sprint.

After having started with the Sprint, the Development Team meets daily to inspect the progress of the work with respect to the planned Sprint Backlog. The **Daily Scrum** is an event time-boxed to 15 min that gives the possibility to each member of the Development Team to report on (1) what she/he has done the day before that helped the Development Team to meet the Sprint Goal; (2) what she/he will do in the next 24 h to meet the goal; and (3) any impediment that could prevent her/him from achieving that objective. Only the Development Team participates in the Daily Scrum. The Scrum Master ensures that the meeting takes place and lasts for no longer than 15 min.

At the end of the Sprint, a **Sprint Review** is held to verify the Increment and to modify the Product Backlog, if needed. The Sprint Review is open to all key stakeholders who collaborate with the Scrum Team in inspecting the Increment. This is a 4-h time-boxed informal meeting (for 1-month Sprints) intended to elicit feedback and foster collaboration. No multimedia presentation is given. Instead, only just the working software is given. The result of the Sprint Review is a revised Product Backlog that defines the probable Product Backlog items for the next Sprint.

Another kind of event prescribed by the framework is the Sprint Retrospective, which is used by the Scrum Team to inspect itself and create a plan for improvements to be enacted during the next Sprint. The Sprint Retrospective occurs after the Sprint Review and prior to the next Sprint Planning. This is a 3-h time-boxed meeting for 1-month Sprints. The purpose of the Sprint Retrospective is to (1) inspect how the last Sprint progressed with regard to people, relationships, processes, and tools; (2) identify and order the major items that went well and suggest potential improvements; and (3) create a plan for implementing improvements into the way the Scrum Team carries out its work. The aim is to improve the development process and the practices to make them more effective and enjoyable for the next Sprint.

13.3 Agile testing practices

13.3.1 Test-Driven Development

Test-Driven Development (TDD), according to the Agile Alliance, "is the craft of producing automated tests for production code, and using that process to drive design and programing. For every bit of functionality, you first develop a test that specifies and validates what the code will do. You then produce exactly as much code as necessary to pass the test. Then you refactor (simplify and clarify) both production code and test code."

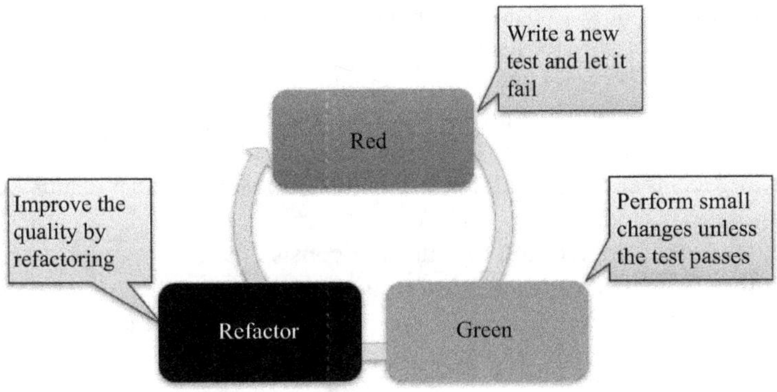

Figure 13.2 The TDD cycle

Software systems built with traditional approaches may exhibit a poor interior quality because they are realized quickly to meet a strict deadline, or unclear requirements, or requirements that may vary during the execution of the development cycle. Consequently, the resulting system may have a large number of defects and be difficult to maintain. With traditional approaches, based on the principle Test-Last, the presence of defects in the code can result from the lack of testing, or from not running tests on software that has shown such defects.

Contrary to traditional approaches, TDD relies on the "Test-First" principle, which lets developers add new code only if a test has failed. The goal of TDD is to get clean code that works.

The Red/Green/Refactor cycle upholds the TDD practice. The cycle (Figure 13.2) is typically executed once for every test and includes the following actions:

1. Write and add a new (unit) test;
2. Make the test run and let it fail. This objective may require new software components (classes, objects, functions, etc.);
3. If the test fails, perform small changes to the software system;
4. Refactor the resulting code;
5. In the case of a new test, restart from the beginning.

There are some misconceptions about TDD, probably due to the word "Test" that stands at the beginning of the name. In reality, TDD is not a testing technique. Instead, it is a design method with the facility of automated tests as a side effect. TDD does not require the writing of all the tests, or a subset of all the tests, nor does it require the coding of the system. Instead, requires that you write one test, write the code to pass that test, refactor, and then repeat.

When building a software system with TDD, developers may need:

- Fixtures: Components used to set the context (and system status) starting from which the test is to be performed;

- Dummy objects: objects passed around but never actually used. They serve, for example, to fill parameter lists;
- Fake objects: objects that have a real implementation. The implementation, however, is "light." As an example, rather than exchanging data with the DB, a fake object uses in-memory data;
- Stubs: objects that provide partial implementations and return results consistent with those expected;
- Spies: stubs that, in addition to returning consistent results, record information useful for the analysis of the testing result;
- Mocks: objects preprogramed with specific expectations. It is a pattern oriented toward Behavioral-Driven Development.

13.3.2 Acceptance Test-Driven Development

Acceptance Test-Driven Development (ATDD) is a technique used to bring customers into the design process. Differently from TDD, which focuses on Unit Tests, ATDD is concerned with Acceptance Tests. Acceptance Tests may be defined effectively starting from User Stories, which are short, very simple descriptions of a feature described orally by the customer or by a user of the system.

The template typically used to pick up a user story is the following: As a <type of use>, I want <some goal>, So that <some reason>.

User stories are often the input information for the definition or the increment of the Product Backlog in a Scrum environment. They are written on index cards or sticky notes and arranged on walls or tables to facilitate planning, transparency, and discussion.

After identifying a new user story, users, testers, and developers define automated Acceptance Criteria that represent the "Done" criteria for the user story. A good template for the description of an Acceptance Criteria is: Given <precondition>, given <actor + action>, then <result>.

Each user story has at least one Acceptance Criteria. From this Acceptance Criteria, an Acceptance Test is derived, which is an instance of the Acceptance Criteria with real data.

As an example, Table 13.1 describes a user story related to an ATM system. The user story describes an operation of cash withdrawal. Two Acceptance Criteria are defined in Table 13.2, i.e., the user story is completed when the system supports these

Table 13.1 Example of a user story

User story	**US1: Cash withdrawal**
As a	Client of the KTL National Bank
I want	to withdraw money from an ATM
So that	I can have some cash

Table 13.2 Example of Acceptance Criteria

Acceptance Criteria	**US1.AC1: Correct withdrawal**
Given	A client of the KTL National Bank with a valid PIN
When	The client requires a cash withdrawal
And	The balance is greater than the amount required
Then	ATM gives money
And	The balance is updated
Acceptance Criteria	**US1.AC2: Card block**
Given	The card holder inserts an invalid PIN
When	The card holder has already made three attempts
Then	ATM retains the card

Table 13.3 Example of Acceptance Tests

Acceptance Test	**US1.AT1: Correct withdrawal**
Initial status	PIN = "123456"; Balance = 10,000 Euro
Input	Cash_required = 300 Euro
Expected result	Balance = 9,700 Euro; Withdrawal = "OK"
Acceptance Test	**US1.AT2: Card block**
initial status	PIN = "123456"; Number_of_previous_failed_attempts = 3
Input	PIN = "123459"
Expected result	Withdrawal = "KO," Exception("CARD_BLOCK")

two scenarios. Formal evidence of the capability of the system is provided when it passes the corresponding Acceptance Tests specified in Table 13.3.

Figure 13.3 shows the ATDD cycle. It starts with the writing of the first user story. Next, the first acceptance criteria is defined and the related acceptance test is developed. From now on, the TDD style may be applied: first, let the Acceptance Test fail; second, small changes are performed in the code until it passes the test; and finally, a refactoring action takes place to improve the quality of the increment. After having passed the Acceptance Test, some Unit Tests may be defined and implemented. When no more Unit Tests are needed, three possibilities may occur: (1) a new Acceptance Criteria may be added to the current user story, (2) a new user story may be identified, or (3) the system is completed.

13.3.3 Behavior-Driven Development

Behavior-Driven Development (BDD) is a development technique defined by Dan North that evolves TDD by shifting the focus from tests to behaviors [47].

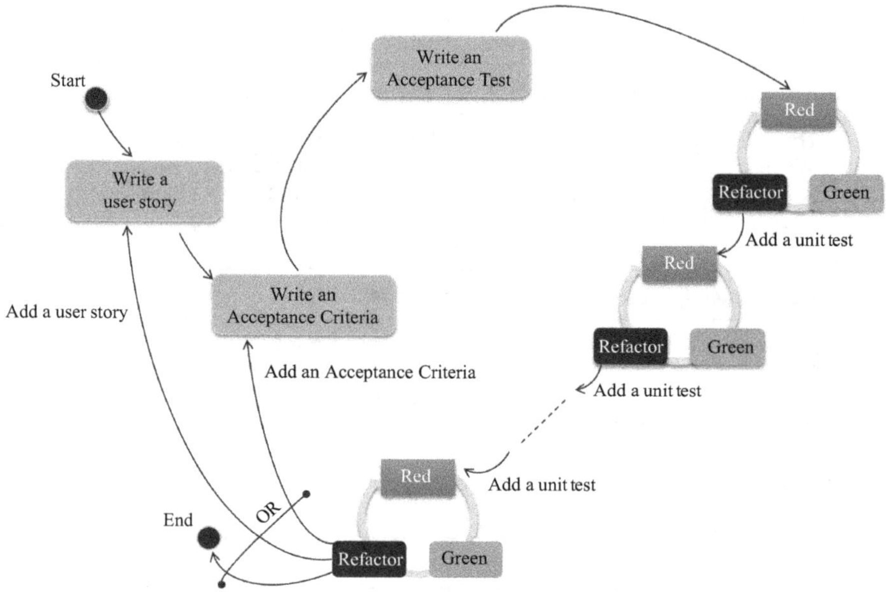

Figure 13.3 The ATDD cycle

BDD combines the general techniques and principles of TDD with ideas from Domain-Driven Design. In BDD, an expected behavior drives the building of new pieces of the system. Expected behaviors are expressed by using the native language in combination with the language of Domain-Driven Design. From this point of view, BDD defines a "ubiquitous language" used by the team in requirements, abstractions (class names, variable names, methods, database tables, columns, etc.), specifications, and documentation. The language is the domain business language. Dan North suggests using phrases for the definition of the tests, starting with the word "should."

For example, considering the given example, one could design a *WithdrawTest* class with tests like the following: *testShoudlFailForMissingPIN()* and *testShould EnsureCashForSufficientBalanceAndCorrectPIN()*.

BDD uses the paradigm Given–When–Then (already described in the previous paragraph) to describe scenarios connected to a user story and to derive, successively, the Acceptance Tests. Moreover, Dan North has developed a tool, namely jBehave, that differs from jUnit—the most popular Unit Test tool adopted by TDD developers in that it allows the construction of tests directly in terms of Given–When–Then constructs. In particular, jBehave provides Java interfaces for the implementation of the statements Given, When, and Then.

13.4 Agile in a regulated environment

Over the years, many industries (including those developing mission critical systems requiring a high degree of safety and dependability) have adopted agile methods,

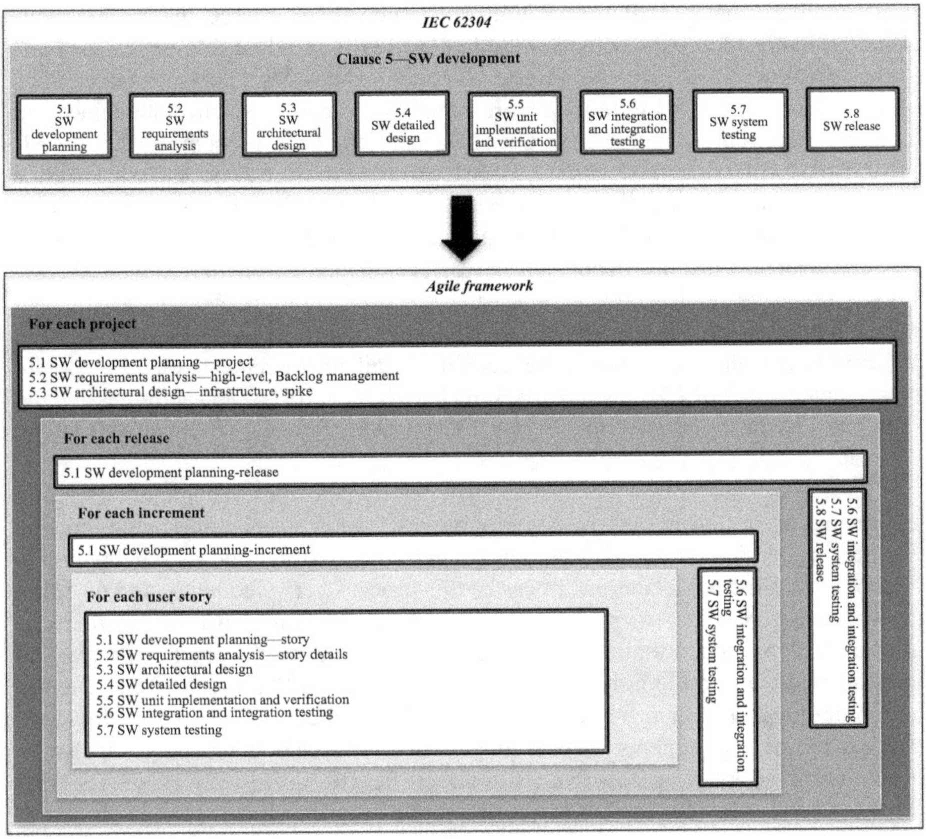

Figure 13.4 IEC 62304—Agile mapping

achieving a high level of quality for their products. However, medical software industries have delayed the widespread adoption of Agile methodologies mainly on account of a few misconceptions. Two of the most incorrect assumptions are (1) regulatory authorities (e.g., the FDA) require the adoption of prescriptive processes like waterfall and V-model and (2) agile methodologies are imprecise, immature, and do not focus on quality.

Concerning the former, it must be clarified that the FDA's quality system does not prescribe any waterfall or V-model based development process. The confusion arises because IEC 62304, which is a standard recognized by the FDA, seems to suggest waterfall as the only accepted process. This is, in fact, not the case because the IEC 62304 standard does indeed require the execution of a certain set of engineering tasks depending on the level of concern for the software under construction, does not encapsulate such tasks in a specific rigid process. As a consequence, any development model that integrates the required tasks may be adopted, even if it differs substantially from waterfall or from V-model, which are, of course, the process models traditionally adopted to comply with the IEC 62304 standard. Moreover, the FDA explicitly cautions against using waterfall for complex devices: "the waterfall models' usefulness in practice is limited, for more complex devices. A concurrent engineering model is more representative." On the other hand, the FDA recognized the "AAMI TIR45: 2012 Guidance on the use of AGILE practices in the development of medical device software" [48] in 2013. The main conclusion of this TIR is that Agile methods allow the achievement of the level of quality demanded by regulatory frameworks and, consequently, can be effectively used to develop medical device software. AAMI TIR45, which covers several key topics such as documentation, evolutionary design, traceability, Verification&Validation (V&V), managing changes and done criteria, is useful to provide an alignment between the regulatory perspective and the Agile perspective. AMI TIR45 shows the mapping between the IEC 62304 activities and the four levels of the Agile methods (Figure 13.4): User Story → Increment → Release → Project.

The latter assumption is a false myth fed by the misconception that Agile approaches are followed by "cowboy programers." After all, planning, requirements engineering, design, and testing are always part of the Agile process. The main difference is that the Agile approach iterates this process very frequently and on small pieces of requirements (increments). The feedback from the customer is received early and often throughout the development of the product. This practice enables the development team to monitor and control the development process more frequently and more constantly than traditional waterfall-based methods. Therefore, it allows teams to produce increments not only much faster, but also with fewer defects than in traditional prescriptive process models. A direct effect of Agile practices, such as User Stories, Unit Tests, TDD, ATDD, and Continuous Integration, is the improvement of software quality constantly throughout the development cycle.

Chapter 14

Project management

14.1 Introduction

In accordance with the definition provided by the Project Management Institute (PMI), "a project is a temporary endeavor undertaken to create a unique product, service, or result."

Every project must create either a completely new or an evolved release of a product, service, or result. Every project must have a certain degree of uncertainty in the result, final quality, time, or cost. Otherwise, it is not a project.

The project has a planned beginning and end, therefore, a definite life-cycle. It may be closed as a consequence of one of the following events: (1) the project's objectives have been achieved; (2) the project's budget (money and/or time) has been consumed, without reaching the objectives, and it is not possible or expedient to increase it; (3) the client requires to terminate the project; (4) the project's objectives can no longer be met; or (5) there is no longer any need for the project.

The PMI has also defined project management as "the application of knowledge, skills, tools, and techniques to project activities to meet the project requirements." It has defined 47 project management processes grouped into five classes:

- **Initiating**: This phase focuses on starting the project. It aims at defining the project at a high level. The feasibility and worthiness of the project is not always certain in this phase. As a consequence, a feasibility study, which begins with a business case, may be necessary. If it is decided to start developing the project, a project charter that outlines the purpose and requirements of the project should be realized. The specification of the project's technical requirements is not due in this phase, rather, it will be formalized successively.
- **Planning**: This phase concerns the definition of a road map for the development of the project, including a definition of the goal of the project. A management plan is detailed, therefore requiring an identification of the scope, costs, timetable, expected quality, and resources necessary.
- **Executing**: This phase of the project management is strictly related to the development life cycle of the product. From the project management point of view, during this phase, progress in the development of the project is represented by the release of a deliverable.
- **Monitoring and controlling**: Monitoring is related to the measuring of the project progression, whereas controlling is related to the identification and execution of

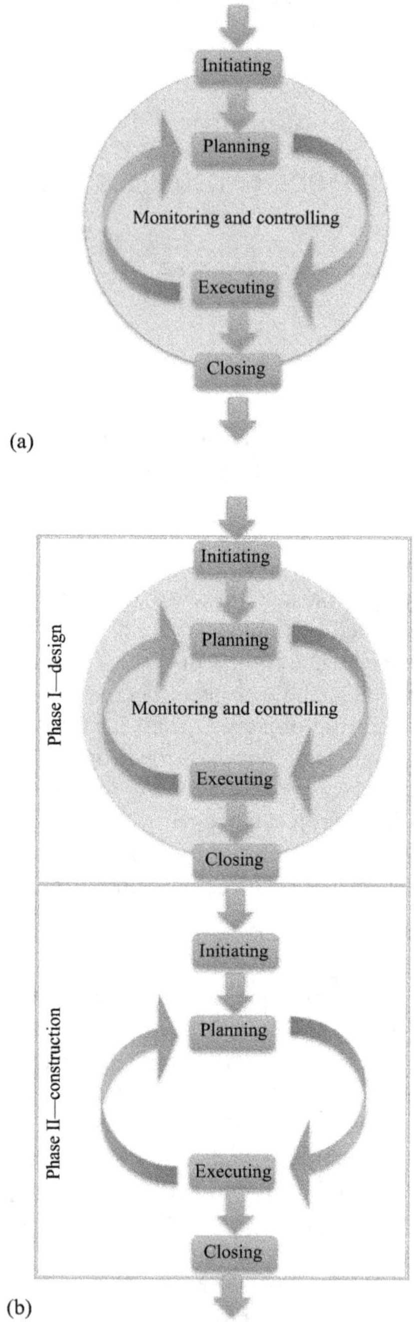

Figure 14.1 Examples of Process Groups Interactions. (a) Example of a single phase project and (b) example of a two-phase project

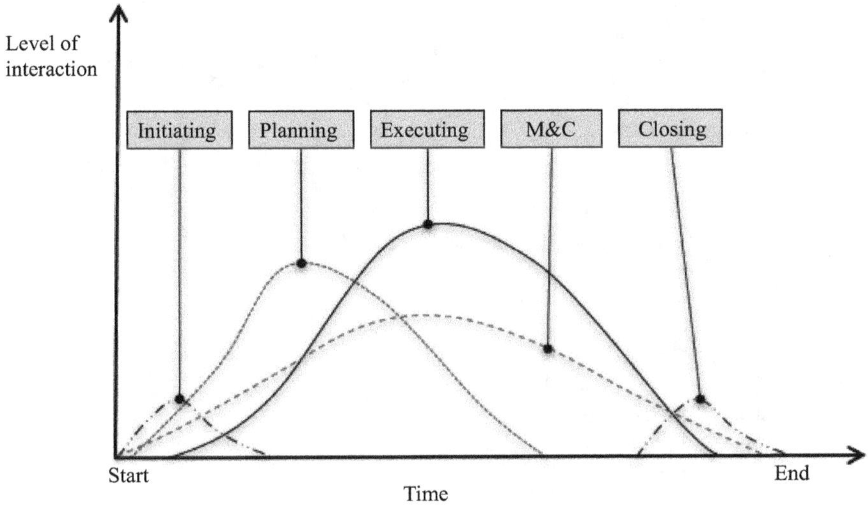

Figure 14.2 Process interaction

corrective actions when deviations from the plans are observed. Key Performance Indicators (KPIs) may be used to determine if the project is on track.

- **Closing**: This phase concerns the closure of the project. The Project Manager, who is the person chosen by the organization to lead the project development team in order to achieve the project objectives, still has some activities to perform, such as archiving deliverable and results, reporting on the execution of the project and learning lessons.

The Process Groups interact with one another as shown in Figure 14.1(a) and (b). The former figure shows the case of a single phase project as, for example, a project for the installation of medical equipment in a hospital department. The latter shows a two-phase project. The example is related to a project that comprises a specific phase for the design and a subsequent phase for the construction. In both cases, with regard to a specific phase, after the Initiating Process Group has completed its activities, a cyclic interaction between the Planning and the Executing Process Groups takes place until the end of the project, which requires that the Closing Process Group fulfills its responsibilities. The level of overlap between these process groups, at various times, is shown in Figure 14.2.

14.2 Initiating

The Initiating phase comprises those activities that are performed to define a new project or a new phase within an existing project. The aim is to obtain the authorization to start the project or the phase. In any case, in which a new project is intended to be

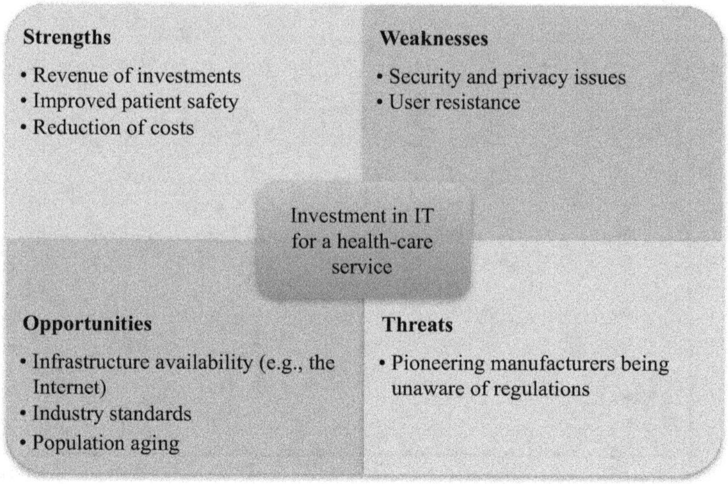

Figure 14.3 An example of a SWOT analysis

realized, it is necessary to perform a careful analysis of the feasibility and worthiness of the project itself. To achieve this aim, a **strengths**, **weaknesses**, **opportunities**, and **threats** (SWOT) analysis may help in deciding whether or not it is expedient to start the project.

A SWOT is an analytical framework whose primary objective is to help any organization to develop a full awareness of the critical factors involved in a decision. There are four types of factors: SWOT. Two of these (the strengths and weaknesses) are internal to the organization, while the other two are external. The two positive factors are represented in the left column of Figure 14.3. The right column reports the negative factors. In the figure, a quite generic SWOT analysis for an investment in IT technologies in a health-care service is reported.

The first step of a SWOT analysis focuses on the assessment of key data. In the health-care domain, this might include the current status of medical technologies, regulations, and standards that apply to the system under construction, sources of funding, intended use, and user needs. In addition to this, the capabilities of the organization are evaluated. In the successive step of the SWOT analysis, the collected data is organized into the four categories of the SWOT matrix. A SWOT matrix is defined with regard to each business option that is under consideration. Finally, the results are reported to the decision-making process as they determine the option that best suits the overall strategic plan.

In a health-care scenario, strengths are factors that have caused organizational performance improvements. Factors that increase health-care costs, as well as those that affect health-care quality, may be weaknesses (e.g., outdated health-care equipment, a poor use of technology, and a lack of financial resources). New business initiatives are opportunities. The development of new health-care programs or increased funds for better health-care informatics is an example of opportunities. Threats, instead, are factors that could reduce the performance of health-care organizations, such as political or economic insecurity, budget deficits, and an increased pressure for the reduction of health-care costs.

14.3 Planning

This phase aims at defining a road map that everyone will follow. This typically includes the following activities: (1) setting the goals; (2) defining the scope; (3) assigning the responsibilities; and (4) planning time and costs.

14.3.1 Setting the goals

The project goals is a statement that describes what an individual, team, or organization hopes to achieve at the end of the project.

One possible tool for the specification of the objectives of a project is S.M.A.R.T. Goal. The origin of this method is uncertain, as well as the acronym, which is expressed in different ways. The following formulation is one of the models most commonly used:

- **(S) Specific**—Generally, when writing about a new business idea, only vague results of the project may be indicated. As the business analysis progresses, it will be possible to declare a more specific goal, which should clearly state what the proponent wants to achieve, why this goal is important, and how he/she intends to accomplish it.
- **(M) Measurable**—The result of the project must be measurable in the sense that it should be possible to assess how far the goals have been achieved. Therefore, a way to measure the progress, as well as the final result, must be defined. A measurable goal should include a plan with milestones and targets that the management team can use to monitor the progress of the project and to establish clearly when the work has been completed.
- **(A) Attainable**—If the goal and the parameters created are not realistic, the project is going to fail. An attainable goal must be realistic and include a plan that breaks the main goal down into subgoals, which are easier to manage and achieve.
- **(R) Relevant**—The relevance of a goal determines the interest in developing a project to achieve it. A relevant goal generally makes sense when measured against the organization's business model, mission, and market.
- **(T) Time-bound**—Every business goal has a time horizon, in the sense that it must be achieved within a defined period of time. Indeed, achieving the business goal outside this time period would not be expedient. For a time-bound goal, a defined time interval, as well as a specific time line for each step of the development process, must be defined.

As an example, we report the following business goals statement for a project, named eHealthPlatform, which we will adopt throughout this chapter.

The SMART goals for the eHealthPlatform project might be established as follows: "We want to realize a platform of open-source software services that will enable the realization of software applications for the (1) medical imaging; (2) telemonitoring of critical conditions; and (3) electronic health record. The early version of such a platform should be ready in 1 year and the final stable version in 2 years from the beginning of the project. We hope to have at least 1,000 downloads within the first year of its availability on the market."

With respect to the SMART structure, we can highlight the following characteristics:

Specific: We want to realize a platform of open-source software services that will enable the rapid realization of applications for the (1) medical imaging; (2) telemonitoring of critical conditions; and (3) managing the electronic health records.

Measurable: We will succeed if the platform receives more than 1,000 downloads in 1 year from its release.

Attainable: The platform is composed of three groups of services, namely those for the medical imaging, tele-monitoring of critical conditions, and managing the electronic health records.

Relevant: The platform will allow the rapid development of a variety of software applications for healthcare.

Time-bound: The early version of such a platform should be ready in 1 year and the final stable version in 2 years from the beginning of the project.

14.3.2 Assigning the responsibilities

Roles and responsibilities concerning the development of the project must be defined. A tool that may be used to accomplish this task is the **RASCI matrix**. Each row of the RASCI matrix is related to a specific activity. Each column reports a role in the project. The responsibility for a specific role with respect to a certain activity is indicated in the cell intersecting the row and column. The acronym RASCI is due to the list of roles, as reported in Table 14.1.

An example of a portion of this chart for the eHealthPlatform is shown in Figure 14.4. It is worth noting that some of the activities/assets reported in the figure are defined at an organizational layer. For example, we may assume that a Quality Policy is defined for any project within the organization. However, at this stage, it is possible to modify the general policy or the related responsibilities in order to satisfy the specific needs of the current project. This is the reason why this task, which belongs to the organizational layer, is reported in the figure at this stage.

14.3.3 Defining the scope

At this stage, a project management plan has been developed and the scope of the project has been clearly defined. The roles and responsibilities are already clear. Therefore, all members of staff involved with the project are aware of the tasks which they will be accountable for.

Table 14.1 Roles for the RASCI chart

Role	Description
Responsible (R)	The person who is responsible for completing the activity. In any case, in which the activity requires the work of many people, the responsible officer is the only person who will ultimately be held accountable if the activity is not completed
Approver (A)	The person who reviews the results of the activity
Supportive (S)	The supportive officer provides the resources to those responsible for completing the activity
Consulted (C)	The people who provides necessary information toward the completion of the activity
Informed (I)	The people who receives information about the progress of the activity

	Chief executive officer	Chief quality officer	Director of customer service	Director of human resources	...	Project manager	Software architect manager	Test manager	Development team	...
Quality policy	A	R	C	I		I	I	I	I	
Project management			I	C		R	I	I		
Risk management	A	AS				C		C		
...										
Communication						SI	I			
Business analysis						SI	I			
Design						SI	R			
Coding						SI	I		R	
Software testing						SI	I	R		
...										

Figure 14.4 An example of an assignment of responsibilities

The Project Manager is responsible for the definition of the scope of the project. This can be achieved by defining the main scope and successively by breaking the project down into manageable sections for the management and development teams.

The **Work Breakdown Structure** (WBS) is a model of an organization of a team's work into manageable sections. The PMI defines it as a "deliverable oriented hierarchical decomposition of the work to be executed by the project team." A WBS allows a specific visualization of the scope into sections that are represented in a multilevel hierarchical structure. Moving from the upper to the lower levels, the amount of details increases and the scope decreases.

Figure 14.5 depicts the WBS for the eHealthPlatform project. The top level is reserved for the project, which is further decomposed into subdeliverables. In general,

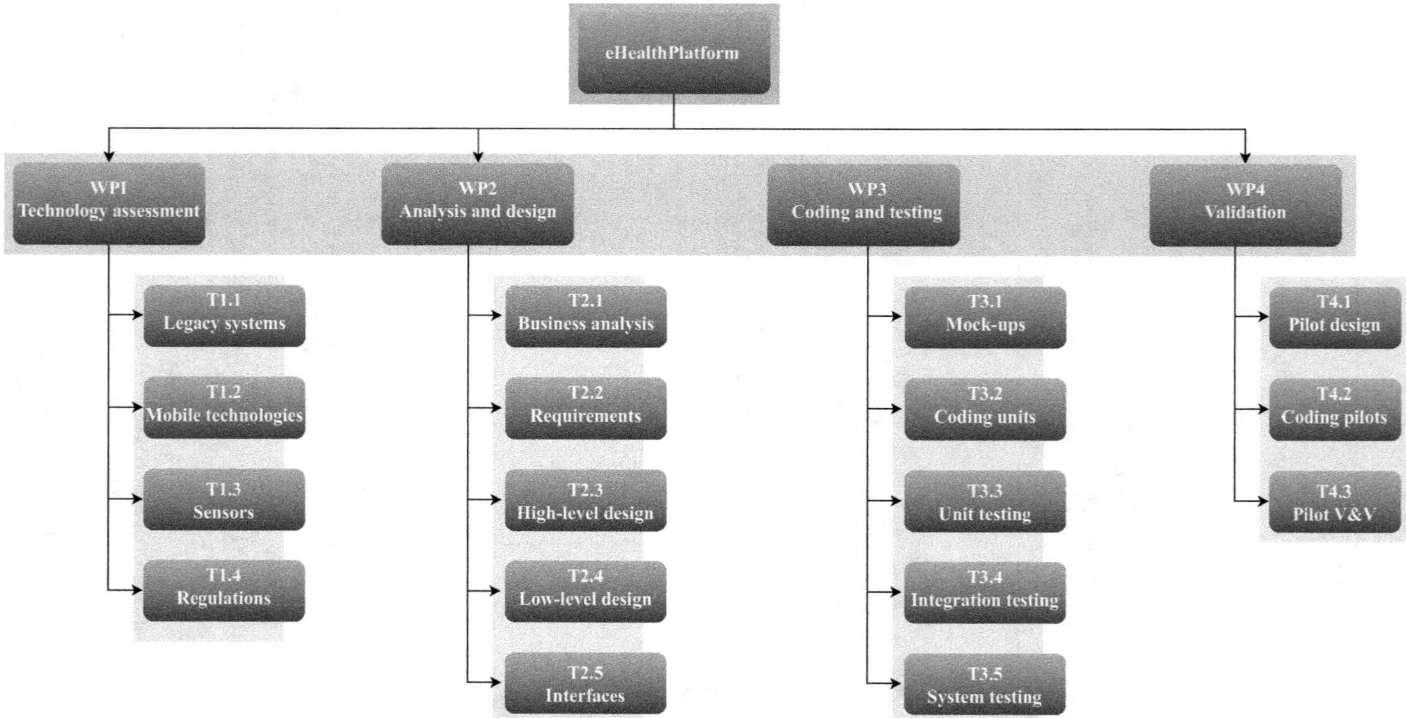

Figure 14.5 An example of a WBS

eHealthPlatform

Month->	Start	End
WP1—Technology assessment		
T1.1 - Legacy systems	1	4
T1.2 - Mobile technologies	3	6
T1.3 - Sensors	3	6
T1.4 - Regulations	1	24
WP2—Analysis and design		
T2.1 - Business analysis	1	8
T2.2 - Requirements	3	10
T2.3 - High-level design	5	12
T2.4 - Low-level design	7	16
T2.5 - Interfaces	5	16
WP3—Coding and testing		
T3.1 - Mock-ups	5	20
T3.2 - Coding units	5	24
T3.3 - Unit testing	7	24
T3.4 - Integration testing	9	24
T3.5 - System testing	11	24
WP4—Validation		
T4.1 - Pilot design	13	20
T4.2 - Coding pilots	17	24
T4.3 - Pilot V&V	21	24

Year I: months 1–12; Year II: months 13–24. M1 (end of month 12)

M1—Milestone: service platform early release

Figure 14.6 An example of a GANTT chart

the hierarchy includes the following levels: the Project (P), Work Package (WP), Task (T), and Subtask (ST), etc.

The Project Manager defines the project WBS by selecting the major functional deliverables and subdividing such deliverables into smaller subdeliverables or systems. These subdeliverables are further decomposed at the next level. Theoretically, the subdivision process may continue until a single person can be assigned to the lowest level deliverable. In practice, a hierarchy of three or four levels is generally sufficient, and the responsibilities for the resulting deliverables are taken on by small/medium development teams.

The adoption of a WBS has a number of benefits in addition to defining the scope and organizing the project work. The project budget, which is allocated to the top levels of the WBS, is rapidly spread among the work packages at the lower level. As the project progresses, any problem affecting a specific activity can be rapidly analyzed in terms of its impact on the expected deliverables or partial results. The WBS can also be analyzed to identify potential risks. If a WBS has a branch, which is not well defined, this represents a scope definition risk.

14.3.4 Planning time and costs

Traditional project management techniques plan time constraints as precisely as possible in advance. The project timetable can be effectively defined by means of a **GANTT chart**, which reports the temporal extension of each activity and task defined in the WBS. An example of a GANTT chart for the eHealthPlatform project is depicted in Figure 14.6. Concerning such a planning activity, some considerations may be useful:

1. Although some tasks are dependent on one another, they overlap. This is the case, for example, in respect of work packages WP2 and WP3. At first glance, the coding and testing activities (WP3) need the result of the analysis and design (WP2). This kind of constraint must certainly hold for a pure waterfall development process. This is not the case of the current GANTT because an evolutionary development process has been adopted. Therefore, an activity can be launched although the required results from the previous activities are not yet stable. On the contrary, for any case, in which an activity definitely needs the final result of another, within the GANTT chart, this sort of dependency may be specified by means of an arrow connecting the end of the previous activity with the beginning of the successive one. In general, four types of dependencies may occur between two activities, A1 and A2: A1 must start before the start of A2, A1 must start before the end of A2, A1 must end before the start of A2, and A1 must end before the end of A2. All such constraints can be represented on the GANTT chart with arrows connecting the starting/ending point of A1 with the related starting/ending point of A2.
2. In the current GANTT chart, a milestone (M1) has been inserted to indicate that, at the end of the first year of the project, the early software platform release is due. This is a critical step requiring that the development of the project should be planned with the possibility of an early verification of the technologies developed, with the aim of improving the results by exploiting the feedback obtained from

such a verification before releasing the product on the market. In general, a **milestone** represents an event particularly relevant for the achievement of the project's goals.

3. Some professional tools allow the Project Manager to indicate the human resources assigned to the project tasks. This is not the case in respect of the current GANTT because the development teams for the eHealthPlatform are quite large.

When close dependencies exist between the tasks of a project, a **Critical Path Analysis** (CPA) may be useful.

A CPA is realized by using an oriented graph in which nodes represent events such as the start and end of a task, and an arrow (arc) between two nodes shows the activity needed to complete that task. The task name, along with the task duration, is written near the arc.

In the GANTT chart reported in Figure 14.7(a), several dependencies are specified. For example, task 5 (coding) cannot start until tasks 2 (development tools selection) and 4 (design) are completed. Figure 14.7(b) reports the equivalent reticular representation. Both charts have been obtained by using a professional tool. Figure 14.7(c), instead, depicts the CPA for the same project. All such charts show the **Critical Path** (in red), which is the sequence of tasks that determines the duration of the project. Indeed, a delay in any of these tasks will imply an equivalent delay in the entire project. This is not the case in respect of task 2 (development tools selection). Indeed, any delay in such an activity, of up to 6 weeks, will not affect the duration of the entire project.

A variation of CPA is represented by the **Program Evaluation and Review Technique**, which takes into account the uncertainty with respect to task durations. In particular, with this kind of analysis, the Project Manager has to estimate the optimistic time (o), the most likely or normal time (m), and the pessimistic time (p). The expected time (te) is computed using the formula:

$$\mathrm{te} = \frac{o + 4m + p}{6} \tag{14.1}$$

Concerning the project cost planning, it helps Project Managers to baseline the overall project budget and the project sponsors to guide the project cost–benefit analysis. This represents an attempt to identify the cost elements to be consumed during the project life cycle, for example, in terms of effort, materials, equipment, and consultancies, etc.

In traditional project management approaches, cost planning follows the time planning. As a consequence, once the Project Manager knows the duration and nature of tasks, she/he can focus on the resources, equipment, consultancies, materials, and other assets needed for the completion of such tasks within the planned deadline. Typically, the WBS, along with the key milestones, serves as a reference for the realization of the cost planning.

The most common tool for cost and budget estimation is a data sheet. Figure 14.8 shows an example that reports the costs for human resources, consultancies, and other items.

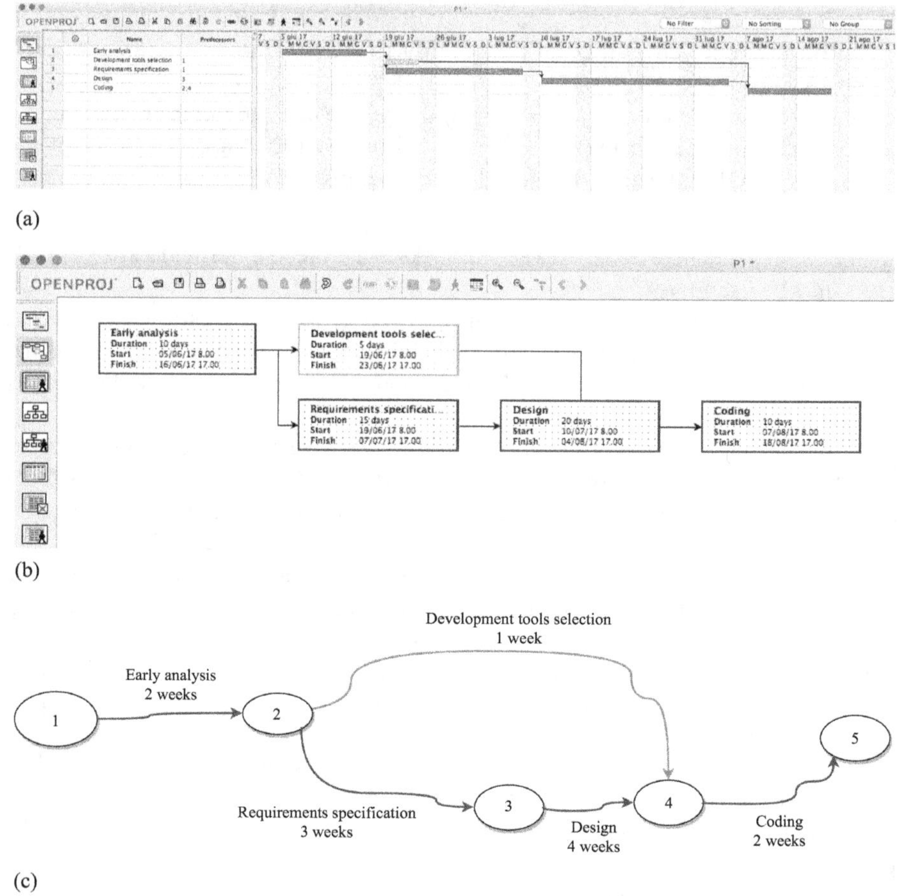

Figure 14.7 (a) Example of dependencies in a GANTT chart, (b) equivalent reticular representation, (c) critical path analysis

14.4 Executing

This is the phase where the project deliverables are realized and released. It does not concern only the development tasks for the project. It also includes status reports, for the realization of meetings, development updates, performance reports, and all activities related to the preparation of such meetings.

A special meeting is the “kick-off” meeting that marks the beginning of the Project Execution phase. During this meeting, all the responsibilities are communicated to the participants.

During this phase, the Project Manager’s role changes. Indeed, if her/his role has so far been focused on planning activities, with the beginning of the Execution Phase, the Project Manager shifts to taking a leading and controlling role.

Internal expenses	Planned budget details			
Salaries from TimeSum	**Hourly rate**	**# Hourly**	**Total cost**	
0	£ 0.00	0	£ 0.00	
0	£ 0.00	0	£ 0.00	
0	£ 0.00	0	£ 0.00	
0	£ 0.00	0	£ 0.00	
0	£ 0.00	0	£ 0.00	
0	£ 0.00	0	£ 0.00	
0	£ 0.00	0	£ 0.00	
Salaries	**Daily rate/Hourly rate**	**# Days/#Hours**	**Total cost**	
Name/Type of resource	£ 0.00	0	£ 0.00	
Name/Type of resource	£ 0.00	0	£ 0.00	
Name/Type of resource	£ 0.00	0	£ 0.00	
Name/Type of resource	£ 0.00	0	£ 0.00	
Name/Type of resource	£ 0.00	0	£ 0.00	
Name/Type of resource	£ 0.00	0	£ 0.00	
Name/Type of resource	£ 0.00	0	£ 0.00	
Other internal expenses			**Total cost**	
Type of expense			£ 0.00	
Type of expense			£ 0.00	
Type of expense			£ 0.00	
Total internal expenses				£ 0.00
External expenses				
Consulting costs			**Total cost**	
Type of consulting			£ 0.00	
Type of consulting			£ 0.00	
Type of consulting			£ 0.00	
Capital expenditures			**Total cost**	
Type of expense			£ 0.00	
Type of expense			£ 0.00	
Type of expense			£ 0.00	
Total external expenses				£ 0.00
Total budget				50.00

Figure 14.8 An example of budget sheet

14.5 Monitoring and controlling

This phase overlaps with the execution phase and concerns measuring the project progression and performance, and monitoring that no remarkable deviation from the project plan is taking place.

Obviously, the GANTT chart and the estimated cost sheet are the input items for this phase and represent the indicators for the monitoring. Any time, a remarkable deviation is detected, a control action must be undertaken. For an example, if at a certain time, it is realized that an activity of the critical path is going to be delayed, the Project Manager could decide to assign more people to the development team until the completion of the activity, in order to avoid or reduce the delay.

In general, Project Managers may use KPIs to determine if the project is on track and how satisfactory the state of progress is. KPIs may be classified into four categories. In the following section, the categories are reported, together with some examples of KPIs:

- **Timeliness**: This is related to the project timetable and deadlines:
  - *On-time completion percentage*: The percentage of tasks completed within the planned deadline;
  - *Time spent*: The amount of time spent on the project by all team members;
  - *Number of adjustments to the schedule*: The number of adjustments performed to the schedule during the execution;
- **Budget**: This is related to the way money is spent and whether or not the costs are under control within the planned budget:
  - *Budget variance*: The variance between the planned budget and the actual budget spent at the closure of the project;
  - *Budget adjustments*: The number of adjustments performed to the budget during the execution of the project;
- **Quality**: This is related to how well the project has progressed:
  - *Customer satisfaction*: The degree of customer satisfaction measured, for example, by a survey;
  - *Number of errors*: The number of artifacts that it has been necessary to reperform during the project. It is related to the amount of work that has had to be redone during the project and has caused budget and calendar revisions;
  - *Unresolved failures*: The number of failures not resolved at the time the product is released;
- **Effectiveness**: This is related to how much effective time and money have been spent for the development of the project:
  - *Return of investment*: The benefit from the project with respect to the investment;
  - *External resources needed*: The amount of external resources needed (e.g., consultancies) because of a lack of internal resources.

14.6 Closing

Once the project is complete, the Project Manager still has certain activities to perform. She/he has to prepare a list of items that were not accomplished during the execution of the project. The final project budget must also be prepared, as well as a project report for the conclusion. Finally, all project documents and deliverables must be collected and archived.

The Project Manager may also organize a final meeting to evaluate successes, identify failures, and establish any lessons to be learned from the project development.

It is worth noting that, in the case of a project that has developed a medical device, the project closure may be postponed until the withdrawal of the product from the market. Indeed, in any case, in which a postmarket surveillance is required, some of the closure activities cannot be accomplished at the end of the development cycle.

Chapter 15
Risk management

15.1 Risk assessment overview

In this chapter, we present a risk management approach based on **Probabilistic Risk Assessment** (PRA). PRA is defined as a systematic and comprehensive risk assessment methodology used to identify and evaluate risks in each life-cycle phase of complex engineered systems across different industry sectors (e.g., chemical, avionics, aerospace, etc.).

The classic PRA approach involves techniques such as fault tree (FT) and event tree (ET), which leverage a Boolean logic to describe the formation of system failures by means of basic events. The accuracy of this approach decreases when the retrieval of probabilistic information about failures is difficult. However, it is still used to compare evaluated risks with predefined risk rankings in order to support improvement during the system design and development phases.

Usually, ET is used to discover and document accident scenarios. As such, it shows the paths that from an initiating event, i.e., a deviation from the proper operation, lead to end states (failure/success) by passing through a sequence of intermediate or pivotal events connected to each other by a cause–effect relationship. On the other hand, FT helps in discovering the cause–effect hierarchical relationship between a top event (effect), usually not observable, and its sub- or basic events (causes).

A major limitation of the FT and ET approaches is the incapability of specifying the repeatability of events as well as the dependency of the occurrence of events on a temporal basis.

Dynamic PRA (DPRA) is useful to address these aforementioned limitations relating to the classic approach. Such a methodology is complementary to the PRA techniques; in that, it is able to explicitly handle interactions between components and process variables with the aim of modeling more realistically the system dynamics to analyze. For example, DPRA approaches offer the key advantage over traditional static ET/FT based PRA methods in that they provide simulations that can represent event sequence timing.

The risk management methodology described in this chapter is compliant with ISO 14971. Figure 15.1 outlines both the main stages and the techniques adopted by this approach, as well as how such phases cover the risk management activities defined by ISO 14971.

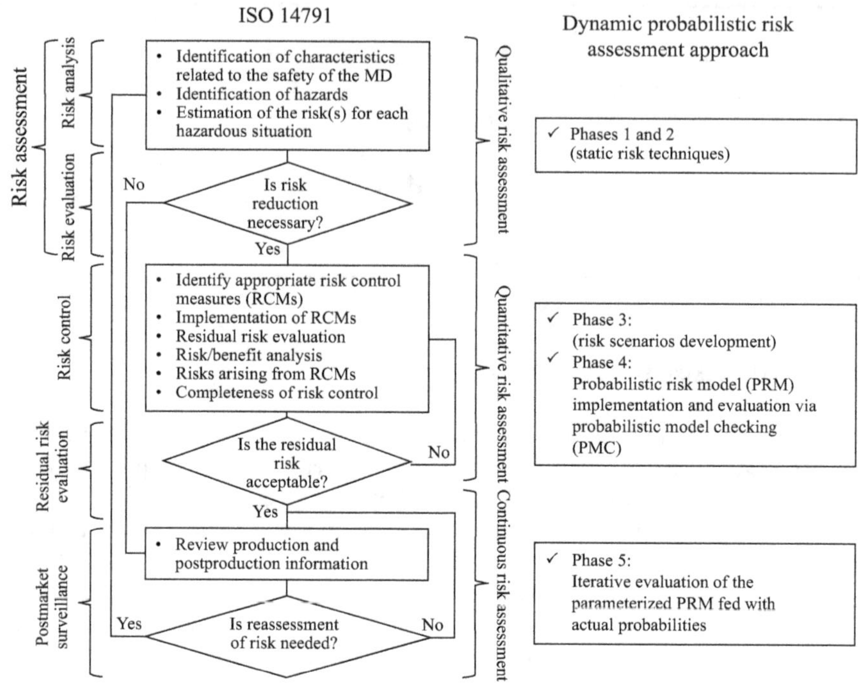

Figure 15.1 A proposed PRA approach compliant with the ISO 14971 standard

The methodology comprises three stages, defined as *Qualitative Risk Assessment*, *Quantitative Risk Assessment*, and *Continuous Risk Assessment* and five steps:

1. Safety critical factor identification
2. Risk analysis
3. Safety risk scenario development
4. Probabilistic risk modeling (PRM)
5. Risk evaluation.

Qualitative Risk Assessment stage includes phases 1 and 2. It aims at recognizing and selecting safety risks by exploiting static RA techniques. This stage is the classic and widespread approach used by manufacturers to conduct risk management for the development of their products. It is the state-of-the-art set of techniques used by most of the Medical Devices industry for low and medium risk devices.

Quantitative Risk Assessment consists of phases 3 and 4. At this stage, risk scenarios are developed by using ET and FT analyses and represented by appropriate PRMs. The aim is to give to the risk analyst a means of investigating the dynamics of how hazardous situations (HSs) develop. ET and FT analyses combined together prove to be a valuable method of, respectively, elaborating risk scenarios i.e., the sequence of higher level (even not observable) events and examining in detail the causal (observable) events composing the higher level ones. Developing risk scenarios also gives the advantage of identifying and documenting the Risk Control Measures (RCMs) taken into account in the design to avoid, reduce, or mitigate the realization of risks. The benefit is to encode the ET/FT analyses into the PRM, e.g., Markovian state transition automata, which enables the risk manager to leverage the efficacy of probabilistic model checking to quantitatively analyze both risks and RCMs by exploring and evaluating different sequences of events.

Continuous Risk Assessment has the objective of providing risk evaluation (phase 5) by parameterizing the transition probabilities of the risk models. Risks can be reevaluated as soon as fresh data are collected and transition probabilities related to the model parameters are computed. This may be particularly useful for the postmarket surveillance, when new data can be collected in the field. A possible way of performing Continuous Risk Assessment is by model checking the PRMs realized at the previous stage.

Throughout this chapter, we refer to a case study related to the Intelligent System already introduced in Chapter 11, whose aim is to automatize the examination procedures performed within a hospital Department of Nuclear Medicine.

The intended use of the system, describing its principal goals, is:

- to free medical staff from guiding patients within the department in such a way as to reduce their exposure to radioactive agents;
- to manage the examination procedure in order to optimize patient scheduling. Patients are automatically located and guided by using and interacting with these devices: Radio-frequency Identification (RFID) and Personal Digital Assistant (PDA);
- to monitor and handle anomalous situations in order to promptly inform medical staff about any abnormal conditions of patients, devices, and the environment (e.g., leaks of radioactivity agents), and to take under control any HSs. For such functionalities, on the one hand, patients are equipped with a dosimeter so as to monitor the radioactivity absorbed/emitted by their bodies; on the other, medical operators in charge of managing anomalies need to be equipped with a PDA to obtain information about anomalies.

We will now move away from a specific to an abstract nuclear medicine department by assuming it is made up of four locations: an *Acceptance Room* (AR) where the patients are accepted into the department to wait for their examination procedure; an *Injection Room* (IR) where the patients are injected with radiopharmaceuticals, i.e., radioactive agents, according to the diagnostic imaging examinations she/he has to undergo (e.g., a blood volume study, bone scan, brain scan, etc.); a *Waiting Room* WR where the patients wait for the examination after having been injected and until the radiation level reaches the right range; a *Diagnostic Room* (DR) in which the examinations are performed. The last three rooms are equipped with short-range RFID readers so that the patients can be tracked by the system. Inside the AR, an operator registers a patient to be examined and equips her/him with a bracelet containing an

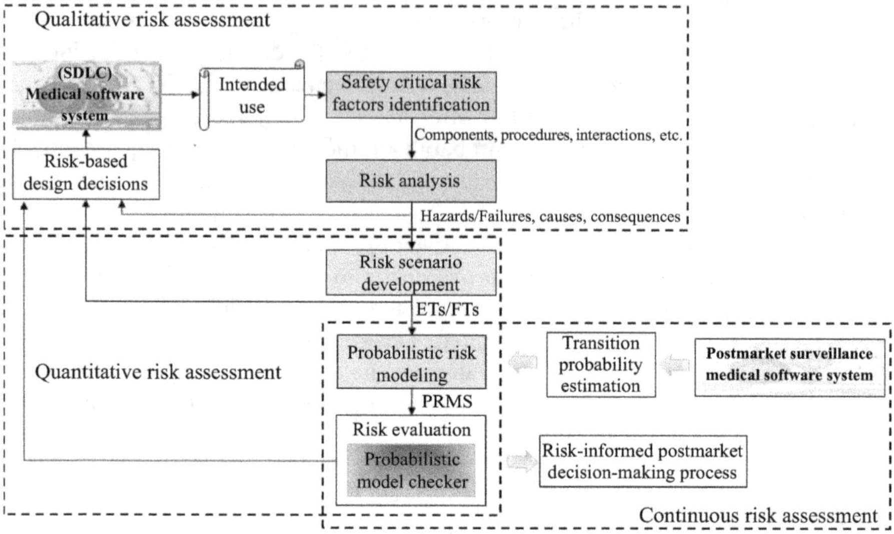

Figure 15.2 A three-phase risk assessment methodology

RFID tag, an accelerometer, a radiation dosimeter, and a PDA. After the registration phase, the system receives data streams from these sensors in such a way that it can monitor, control, and manage the patient's health status and activities, as well as her/his location within the department.

15.2 Risk assessment workflow

Figure 15.2 illustrates the internal details of the stages and phases which constitute the core of the RA methodology. In particular, it is indicated how risk-related information flows through these phases, and the loops needed to feed such information back into the development life cycle to support risk-informed design decisions.

Figure 15.3 highlights the artifacts produced by each phase of the methodology.

Phase 1: Safety critical factor identification. As input, this phase takes the intended use of the system, which is what matters most in the medical device field in terms of the evaluation of safety issues. The goal of this first phase is to identify all the risk factors directly or indirectly linked to the system in which hazards for the safety of humans may hide. To achieve this purpose, a risk factors analysis is carried out with a human-centered approach. By focusing on one user category which the system is intended to interact with, each use case is examined against the procedure/function it realizes, the system components involved, and every physical object of the environment which could be involved in the interaction between the system and the user. The output of such a phase is an enumerated list of risk factors classified for users and use cases, i.e., a risk factor table.

From the intended use of the case study, it is possible to identify two human actors (categories): patients and medical staff. By focusing on use cases related

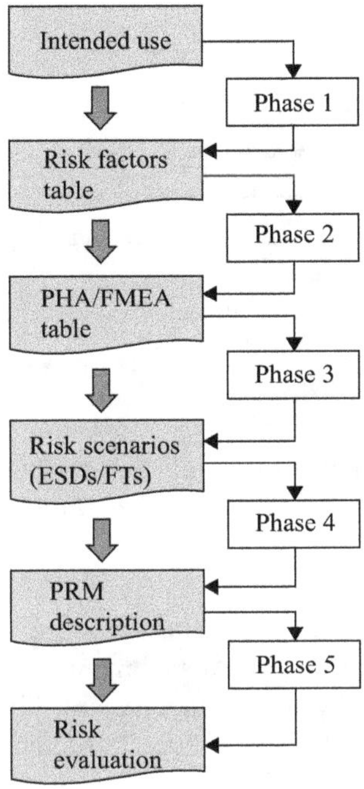

Figure 15.3 Document flow through the phases

to such actors, we can identify the related safety critical factors. For instance, by selecting the use case relating to the guidance of patients through the activities of the examination procedure, critical factors can be identified in those system components, functionalities, or interactions with the actor which could lead the patient to fail in performing the examination procedure. In our case, for instance, the critical factors are the Wi-Fi connection of the PDA to the software system, the PDA batteries, and the fact that the patient can separate her/himself from the PDA because it is detachable (a more detailed analysis of the case study is reported in Section 15.5).

Phase 2: Risk analysis. The second phase comprises two qualitative risk assessment methods, namely PHA and Failure Mode Effects Analysis (FMEA). In the early stage of the system development, when the design is still not available, the former method helps in identifying and binding together the hazards, causes and consequences related to each critical factor identified in the previous phase. For this reason, the risk factor table—the artifact of phase 1—is the input that in phase 2 guides the discovery of safety risks. The latter technique is generally used to establish the failure modes of the system components, and, therefore, it can be applied as soon as a first attempt at the system design is ready. The output of this phase is a PHA/FMEA table which lists hazards, causes, and consequences associated to each risk factor.

For the case study, by taking the critical factors mentioned in the description of the previous phase, a PHA approach may be used to derive various risks

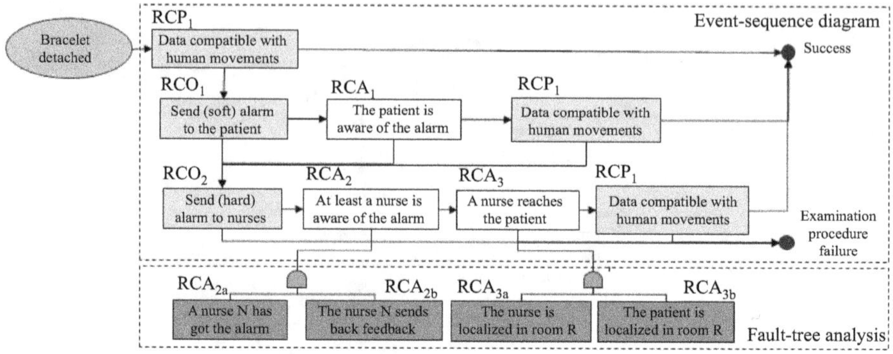

Figure 15.4 Example of an event-tree/fault-tree analysis to represent a risk scenario for the Nuclear Medicine Management System

which are independent of the internal design of the system, for example, a Wi-Fi malfunction, drained batteries, or a detachment of the sensors. After any design, decisions have been taken by, for instance, establishing the Wi-Fi configuration or the type of power supply subsystem of the PDA, the FMEA is then used. This can serve to bring out the hazard as the signal interference, a possible cause identified in the kind of Wi-Fi band used, and the consequence with respect to the associated use case and/or actor, e.g., a loss of connection leading to the possibility of the patient not being monitored/guided in following her/his examination procedure.

Phase 3: Safety risk scenario development. The third phase receives as input the PHA/FMEA table of the previous phase. The purpose of this phase is to analyze and develop risk scenarios, i.e., the sequences of events through which, from hazard causal events (e.g., system failure, human error, etc.) a HS having negative consequences for the patient may occur. An important benefit of this phase is the elaboration of RCMs to identify and interweave into the risk scenario so as to prevent the occurrence of HSs or to reduce the negative consequences.

To represent risk scenarios we adopt the widespread combination of ET/FT. An ET is developed by analyzing the causal–effect chain which connects the initiating event (the perturbation with respect to normal conditions) to the HS (i.e., the end state).

As an example, Figure 15.4 reports an ET/FT analysis conducted to determine a risk scenario related to the detachment of the bracelet from the patient's wrist, which means that the system would not be able any more to locate and guide the patient during the examination procedure. Each square of the ET represents an intermediate event to be evaluated. In particular, each event could be seen as a Boolean predicate which provides a decision point: if it is evaluated to be true, the next event to evaluate is reached horizontally from left to right in the scenario; otherwise (i.e., it is evaluated to be false) the next event is reached vertically from the top downwards. So, the underlying execution model is sequential. Pathways toward the "Success" end state represent a successful management of the HS to be prevented; on the contrary, pathways leading to the "Examination procedure failure" end state represent dangerous (unsuccessful) sequences of events.

If the ET contains intermediate events which prove to be too abstract, i.e., not directly observable and measurable, the deductive reasoning provided by the FT analysis is exploited to discover the more basic casual events realizing the abstract one. For instance, in Figure 15.4 the intermediate events RCA_2 (risk control action-2) and RCA_3 could be considered not measurable events which, therefore, require to be broken down by an FT analysis to discover their basic events. Once the FTs are constructed, each event of the ET could be expressed in terms of basic (observable) events. So, for example, RCA_2 is seen as the logic AND between RCA_{2a} and RCA_{2b}.

The ET/FT analysis has some limitations. First, the order of the event sequence is fixed, i.e., loops are not possible. Second, the temporal aspect is not considered within the model, so it is not possible to evaluate the variation of risk over time. Third, the Boolean logic enables the risk analyst to encode the predicate associated with whether the event has happened or not, but does not permit a description of the case in which the event is not delivered at all, i.e., the RCM associated with the event does not perform as expected. However, since ET/FT is a common and well-known formalism by which system designers, risk analysts, and domain experts can communicate to identify those safety risk scenarios to be taken into account, it is still useful to exploit ET/FT as a tool for conducting a risk analysis as well as for producing evidence for the certification process. The ET/FT risk scenarios that need to be developed form the artifact of this phase.

Phase 4: Probabilistic risk modeling. To overcome the limitations of ET/FT mentioned in the previous phase, the fourth phase foresees the instantiation of a (Markov-based) PRM. This is a formal model that represents the stochastic dynamics of hazard scenarios so as to be able to perform a time-dependent risk assessment. As input this phase takes the ET/FT risk scenarios from which the PRMs are built and given as output.

Phase 5: Risk evaluation. This phase deals with evaluating the PRM risk scenarios. Since PRM is a Markov-based model, a probabilistic model checking technique may be exploited as an automatic, efficient, and powerful solution to perform quantitative analysis and evaluation on the developed risk scenarios. As input, this phase takes the PRMs developed in the previous phase, and a set of qualitative and quantitative properties (expressed in an appropriate temporal logic) to be evaluated.

An example of phases 4 and 5 is shown later in this chapter. There, a formal definition of PRM, an explanation of how it is built from ET/FT risk scenarios, and an evaluation by performing probabilistic model checking is provided.

As a further step, PRMs may be instantiated as parametrized models whose parameters are those transition probabilities of interest among the risk scenario states. In doing so, two important benefits are obtained. First, during the system development life cycle, it is possible to address the uncertainty of transition probability estimations, as well as to conduct a sensitivity analysis on the risk evaluation with respect to any variation of the parameter values within exploratory ranges. Second, after the deployment of the system, manufacturers can use the latest and most reliable version of the representative PRMs in a continuous risk assessment process as a means of monitoring the system during postmarket surveillance. This is achievable by feeding the actual transition probabilities computed on real postmarket information into the PRMs so as to continuously evaluate and monitor the safety and effectiveness of the system.

It is worth noting that, as Figure 15.2 shows, both qualitative and quantitative information on risks coming from phases 2, 3, and 4 are fed back into the decision-making process. Such information helps to drive the system design during the Software Development Life Cycle, particularly with respect to which aspects prove to be the most critical, therefore requiring a more detailed analysis, as well as verification and validation activities to ensure correctness.

15.3 Static versus dynamic safety risk scenarios

The development of dynamic scenarios follows inductive reasoning in order to identify the sequences of events of interest. Here, an event is used as an abstract concept able to model the state of affairs of the macro-system given by thinking of the integration between the system, the end users, and the environment. A risk scenario takes into account all those elements of such a complex macrosystem (e.g., the system software and hardware components, human behaviors, and environmental factors) which are connected by a cause–effect relationship in the formation of a hazard situation.

In order to develop a risk scenario, it is useful to exploit a "human-machine task analysis approach" as a way of providing a linear description of the observable events. In particular, to help to determine those risk-oriented tasks specific for the system, the following five key principles apply:

1. Take into account those interaction constraints in which hazards for users can be present. This is important, for instance, when the system provides a functionality which requires specific interactions/actions performed by the users in order to succeed. For example, in the case study, the attaching of a wearable sensor to patients performed by the medical staff, or the execution of an action suggested by the system (e.g., moving toward the IR) to be executed by the patients.
2. Formalize those interaction rules for which the system decision-making process used to support the medical activities may experience failures potentially harmful for the users. An example is when a caregiver could be required to respond to an alarming situation (e.g., a patient in a contaminated area) by carrying out predefined emergency actions (locate the patient and take her/him away) within a specific time interval.
3. Consider each user category, e.g., Alzheimer's patients, radiologists, clinicians, etc., and system component involved in the risk scenarios. The intention is to model both the human and device behaviors from a risk perspective. In particular,

model the misbehavior of each user expected to interact with the system (or part of it) so as to take into account user-dependent requirements and interactions. For instance, a patient could be requested only to start/stop a wearable sensor she/he is wearing, whereas a medical operator has to take control and interact with the whole system.

4. Identify those constraints defining the context in which the system and its users have to interact in order to achieve certain some medical objectives. As an example, for the component of the case study which deals with monitoring the patient's position, it is important to know the extent of the area to monitor, its characteristics, whether it is in an open or closed environment, etc.
5. Model those human behaviors and/or actions which could affect the occurrence of a HS. For instance, if the system considers wearable sensors, it is important to model the "turn on/off" and "put on/take off" actions that a patient may carry out.

For the last two points, it is worth emphasizing that the set of human actions varies according to both the category of users, as well as the context in which the user and system interactions take place in order to carry out certain medical procedures or treatment. For instance, patients and caregivers act differently with respect to a medical device used in a health-care environment.

To develop a risk scenario, the risk analyst should consider the following key elements:

- each device component (e.g., a wearable and/or environmental sensor, or medical device) which is interconnected to constitute the system as a whole;
- the set of end user categories (e.g., patient, caregiver, doctor), and for each one its set of performable actions; and
- a set of user-system–environment interaction constraints.

The ET/FT analysis has been successfully used in the aerospace industry. The former technique is used to describe the system behavior in the case of anomalies, while the latter is a way of providing a methodical building of a causal model so as to identify and quantify the contributing causes of the events in the ET. The idea is to continue to use this effective analysis to develop hybrid risk scenarios in which both the system and the human behaviors are interwoven to cope with the ongoing HS. In the methodology, however, the ET/FT combination is exploited as a higher level industry-oriented formalism to develop, explore, refine, and document safety risk scenarios, but not to conduct a risk evaluation. By so doing, we have the advantage that Medical Device designers can use the ET/FT artifacts as objective evidence related to safety risk which is of great interest to the Regulatory Agencies when the certification process is carried out. This higher level representation is then exploited as a basis to encode risk scenarios into a lower level formal representation based on Markov models, by means of which the analyst conducts a thorough qualitative and quantitative risk assessment.

The starting point in developing dynamic risk scenarios in this PRA methodology is given by the basis information resulting from the risk analysis phase, i.e., HSs/failures, causes, and consequences (harms). Causes are seen as initiating events, consequences are associated with (possibly harmful) end states, whereas HSs are mapped as unsafe states. The analysis and development of a risk scenario takes advantage of such information to allow the exploration and identification of alternative/complementary RCMs to be activated to address predetermined anomalies.

It is worth observing that the evaluation of risks for Medical Devices and Medical Device Software is strongly dependent on the related human factors, and characterized by the way different categories of users interact with both the system and the environment. From such a consideration, the development of the risk scenario is supported by distinguishing RCMs in:

Risk Control Operations (RCOs) performed by the system device components;
RCAs carried out by the end users and supported/monitored by the system;
Risk Control Points (RCPs), which are certain points in which information about the state of the system, the users, or the environment needs to be checked in order to take a decision about the course of the scenario.

For example, referring to Figure 15.4, "Bracelet detached" is considered as an initiating event which could result in a HS for the patient leading to a failure in performing the examination procedure. To prevent leading to a failure such a risky situation, during the risk scenario development a set of RCMs has been conceived and analyzed to include in the system. In particular, the RCP ($\mathtt{RCP}_1$) executed by the system is designed to check if the data received by the bracelet accelerometer is compatible with human movements or, on the contrary, if it reveals that the bracelet is motionless. In the latter case, the scenario develops by considering two RCOs, i.e., $\mathtt{RCO}_1$ and $\mathtt{RCO}_2$, which implement two alarm levels: the first directed to the patient's PDA, and the second to the nurses' PDAs. In both cases, the ET analysis brings out *abstract events*, i.e., conditions not directly observable or measurable by real events collected by the system. In the example, such an abstract event consists in the patient or nurse being aware of the alarm. By exploiting an FT analysis, instead, analysts are able to consider such an event, which is considered as the top event, in greater detail in order to discover the subevents, or basic (detectable) events, leading up to it. The application of the FT analysis is mandatory for abstract events, since we need to encode our risk scenario and quantify it. For the sake of brevity, Figure 15.4 shows only the FT analysis related to the $\mathtt{RCA}_2$ and $\mathtt{RCA}_3$, but, similarly to $\mathtt{RCA}_2$, it is the FT of $\mathtt{RCA}_1$. For instance, $\mathtt{RCA}_2$ is decomposed into a logical AND operator of two detectable RCAs, i.e., $\mathtt{RCA}_{2a}$ and $\mathtt{RCA}_{2b}$, i.e., "a Nurse N has received the alarm," i.e., at least one PDA owned by one nurse has received the alarm sent by the system, and "Nurse N sends back feedback," i.e., the same nurse who has received the alarm sends back feedback as a confirmation.

Two important aspects left out of traditional PRA techniques are captured in the risk model in this DPRA methodology. Specifically, the time elapsing to reach a state within the risk scenario and the behavior of each human category interacting with the system. Furthermore, the ET structure has the limitation of describing only sequentially scenarios in which events occur, which is highly unrealistic for system since multiple autonomous components and users are involved.

It is possible to progress from the ET/FT scenario blueprint to a discrete-time, nondeterministic, PRM in which a model checking technique can help to explore and evaluate all possibly risky sequences of events. In the following sections, a formal definition of a PRM, and a description of how ET/FT scenarios are formally encoded in a PRM, is given.

15.4 Probabilistic risk model

PRM relies on a formal model to capture human and machine interactions. In the model described next it is possible to represent the stochastic (risk-oriented) behavior of

- the system as a whole (seen as the process controller);
- each autonomous system component (e.g., an environmental/wearable sensor and/or medical device); and
- each user category (e.g., a patient, or doctor, etc.).

The internal dynamics of each device is modeled as a probabilistic state-transition automata, in particular a discrete-time Markov chain (DTMC), in which the next state is selected by following a probabilistic distribution function over all the next states. Since humans are characterized by making choices among different actions to perform in a nondeterministic fashion, it is possible to model human behavior by means of Markov Decision Processes (MDP), a generalization of a DTMC which allows the modeling of nondeterministic strategy policies of actions.

By denoting with U the set of user categories,[1] i.e., $U = \{u_1, \ldots, u_n\}$, and with D the set of device components constituting the system, i.e., $D = \{d_1, \ldots, d_m\}$, we thus formally define a PRM for a system as follows:

$$\mathrm{PRM} = \langle \{\mathrm{MC}_{u_i}\}_{u_i \in U}, \{\mathrm{MC}_{d_j}\}_{d_j \in D}, \mathrm{Comm} \rangle \tag{15.1}$$

where MC_{u_i} and MC_{d_j} are, respectively, the MDP modeling the user u_i and the DTMC modeling the device d_j, which are considered in the risk scenario associated with this PRM.

Comm is the set of the triplet $\{(d_i, sv, d_j)\}$ where sv is the state variable on which the DTMC states of the d_ith device are defined, and which is made observable, i.e., readable, by the DTMC of the d_jth device.

MPD is defined as the tuple $\langle S, s_0, A, P, R \rangle$, where S is the state space defined on a set of atomic propositions, s_0 is the initial state, A is the set of all actions, P is the transition probability among states, and R is the reward function which associates a numerical value to each transition and/or state. A DTMC is defined similarly. Both models represent the dynamics of state transitions based on a discretization of time. To be consistent, the PRM model has to treat time uniformly by considering a time unit for each DTMC/MDP. Since MDPs are defined as discrete-time models whose transitions take place at discrete time intervals, by leveraging PRMs to represent risk

[1]Without loss of generality, a user can represent either an individual, or a category of individuals, e.g., patients having the same pathology.

scenarios, a probabilistic model-checking can be performed to assess two quantities particularly important from the risk perspective:

1. the probability of reaching a state within a T time unit and
2. the maximum/minimum value of T time units by means of which the probability of reaching a state is below/above a certain threshold of interest.

It is important to note that probabilistic model checking, and the associated temporal logics used to formalize properties of interest for analysts, is not restricted to expressing and computing only these aforementioned factors.

15.5 Application to the case study

This section shows the workflow and application of certain techniques to the risk management of the case study introduced in Section 15.1. The analysis has been focused only on a specific, yet representative and realistic, set of risk scenarios useful for the purposes of this chapter. The aim of this section is to report and describe in more detail the application of each phase of the methodology.

15.5.1 Safety critical factor identification

The intended use described in Section 15.1 is the starting point from which the risk analyst identifies two human actors or user categories i.e., medical staff and patients, the system device components that directly interact with each user category and the use cases provided by the system.

Table 15.1 reports some of the safety critical factors identified with respect to the human actors on whom possible HSs may have negative consequences. The factors are grouped by use case in order to monitor and manage the risk assessment process against the system functionalities.

The first two use cases allow the system to automatically manage the identification, acknowledgment, and examination procedure for all patients within the department. The "Patient Monitoring vital signs" and "Anomaly Detection and Control Measures" aim at providing the functionalities necessary for the purpose of detecting, and reducing or mitigating risks, for both patients and medical staff. To be realized, a use case requires that certain constraints and/or conditions are satisfied, and needs to exploit a subset of hardware/software system components. By analyzing such elements, we have determined a list of critical factors. For instance, in respect of the use case "Patient Identification," the critical factor CF1 takes into account that, to be identified everywhere in the department, patients need to pass through a preinstallation procedure of acquiring the necessary equipment (i.e., the bracelet and PDA) as the intended use specifies. This is considered a critical factor since, in fact, any issue that could come up during this phase (e.g., an incorrect association of the patient with the RFID and PDA provided to her/him) may cause HSs to develop. For the same use case, CF2 is a factor related to a system component, i.e., the RFID tag, which is responsible for performing the identification and localization of the staff and patients when the system is in operation.

Table 15.1 Safety critical risk factors

Human actor	Use case	Critical factors
Patient	Patient identification	**CF1.** Equipment installation procedure for patient **CF2.** Identification procedure via (wearable) RFID sensor
Patient	Patient procedure guidance	**CF3.** Patient must follow the examination procedure guidance provided via the equipped PDA **CF4.** The PDA is connected via Wi-Fi to the system **CF5.** The PDA is detachable **CF6.** The PDA is powered by batteries
Patient	Patient monitoring of vital signs	**CF7.** The wearable sensors are connected via Bluetooth to the system controller **CF8.** The wearable sensors are detachable **CF9.** The wearable sensors are powered by batteries
Patient	Anomaly detection and control measures	**CF10.** Patient awareness of warnings rose due to misconduct of the examination procedure **CF11.** Patient preparedness for hazardous situations and/or emergency operations **CF12.** Areas in which patients are forbidden to go to or into
Medical staff	Anomaly detection and control measures	**CF13.** Medical staff must be equipped with a PDA, RFID, and dosimeter **CF14.** Medical staff awareness of alarms related to hazardous situations regarding any patient **CF15.** Medical staff availability to take part in corrective/ mitigation emergency actions

The advantage of this phase is to focus the attention of analysts on the human-centered risk factors, so as to drive all other phases in the analysis of issues and production of artifacts oriented to support the regulatory approval process. In this way, the analyst filters out risks that do not impact on human health, or which do not pertain to core functionalities, e.g., messaging services among doctors, or the reporting of general (nonmedical) data, etc.

15.5.2 Risk analysis

Based on the output of the previous phase, the PHA and FMEA can identify potential hazards, the main causes, the consequences for people, and some preliminary countermeasures. Particularly, the analysis conducted on CF3, CF4, CF5, and CF6 is reported. The hazard lists (reported in a reduced version) which we have obtained are as follows:

H3.1 The patient moves unconsciously into a radioactive area
H3.2 The patient is incorrectly located

H3.3 An uncontaminated patient meets a contaminated one, or goes to/into a radioactive area
H3.4 The medical staff member experiences some complications in injecting the radiopharmaceutical into the patient's body.
H3.5 The radiopharmaceutical diffusion in the patient's body is slower/faster than expected
H4.1 PDA malfunction
H4.2 Wi-Fi Network malfunction
H4.3 System controller malfunction
H5.1 The PDA strap is damaged
H5.2 The patient is distracted
H6.1 Low battery charge

Each list reports the identified high-level hazards Hn.x, each related to the critical factor CFn. Most of such hazards are related to system components and functionalities such as H3.2, H4.x, H5.1, and H6.1, whereas the others—resulting from the FMEA analysis—are specific to the NM domain.

Table 15.2 reports a safety-oriented excerpt of the risk analysis conducted in the case study. Each row of the table identifies information which links hazards, causes, HSs, and consequent harms, and also provides some identified RCMs.

These first two phases allow the analysts to identify factors which can go wrong, which system components and functionalities are involved, and the RCM from which can be obtained benefits in terms of reducing HSs or of preventing them from happening.

15.5.3 Risk scenario development

On the basis of the information provided in the previous phase, detailed risk scenarios are conceived as an ET. We focus on RS1 (see Table 15.2) to show the development of a risk scenario. This choice allows us to illustrate a risk model feature taking into account human behaviors, as well as the importance of the temporal aspect is a key factor to consider so as to evaluate risk scenarios with respect to different time lapses.

To develop a risk scenario for RS1, we conducted a human-machine task analysis driven by the associated countermeasures identified in the previous phase. By applying this analysis, it was possible to determine the following system operations and human actions preventing the system from reaching the HS associated with RS1:

O1.1: [system controller] sends a warning to the patient's device
O1.2: [system controller] sends an alarm to the medical staff's devices
O1.3: [system controller] sends an order to the medical staff to solicit the patient to move away
O1.4: [system controller] checks if the patient's radioactivity is lower than a safe threshold
A1.1: [Patient] is aware of the warning
A1.2: [Patient] moves away from the radioactive area
A1.3: [Caregiver] is aware of the alarm

Table 15.2 Risk analysis based on safety-oriented PHA and FMEA

ID	Hazard	Cause	Hazardous situation(Harm)	Risk control measures
HS1	The patient is in a radioactive area (H3.1)	The patient does not follow the guidance provided via PDA	The patient is exposed to radioactivity (patient contaminated)	The patient is warned via PDA; an alarm is sent to the medical staff
HS2	The patient is located in a wrong area (H3.1)	The RFID tag is detached from the patient's body	The patient fails to follow the examination procedure (patient contaminated)	An alarm is sent to the medical staff; the patient is located and provided with another RFID tag
HS3	The DR is still busy when another patient is ready to be examined (H3.1)	Examination delay; accelerated diffusion of the contrast agent	The patient fails to follow the examination procedure (patient contaminated)	The patient is transferred into the HWR for a safe decay of the radioactivity
HS4	Patient localization error (H3.2)	The RFID tag/reader is out of order/range	The patient fails to follow the examination procedure (patient contaminated)	The RFID tag/reader is repaired/replaced or a caregiver is located to track the presence of the patient via her/his PDA
HS5	Patient/medical staff in contact with a radioactive patient (H6.3)	The radioactive control subsystem has failed	Patients/medical staff are exposed to radioactivity (patients/medical staff contaminated)	Any contaminated people are isolated in a decontamination room

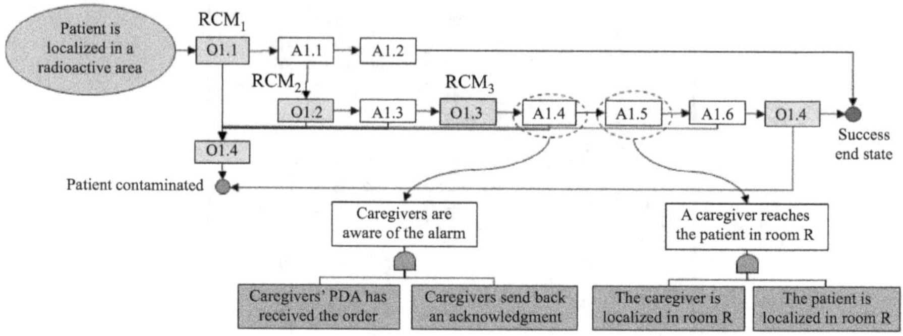

Figure 15.5 Event-tree/fault-tree of the risk scenario RS1

A1.4: [Caregiver] accepts the order
A1.5: [Caregiver] reaches the patient in the radioactive area
A1.6: [Caregiver/Patient] moves away from the radioactive area

Figure 15.5 illustrates the ET representing the sequence of events of such preventive/corrective operations/actions that lead to either a successful or unsuccessful end state (patient contaminated). Furthermore, an FT analysis conducted on some abstract events is shown. The tree structure helps in identifying the real basic events which realize the abstract event.

This phase, usually not taken into account by MD manufacturers, has the advantage of bringing out "how" wrong events may happen, and how RCMs can be arranged to accomplish their tasks. The drawback of an ET/FT analysis is that it does not consider the temporal aspect in that it fixes the possible sequences of events to those explicitly determined by the analysts.

15.5.4 Probabilistic risk model

A PRM is defined with respect to a risk scenario and its associated information. Considering RS1, Figure 15.5 shows its development as well as the information drawn from the risk-analysis phases. The corresponding PRM is devised by modeling the behavior of the system controller, of each independent sensor/device, and of each patient and caregiver category. In particular, the model reports only those states and transitions relevant for the risk assessment process.

Figure 15.6 sketches part of the state-transition MDP associated with the system-controller, whose states are the following:

MDP-system-controller states: `safe` (initial state), `patient_in_radioactive_area`, `unsafe` (end state), `<operation>_sent`, `fail_to_send_<operation>`, `ack_<operation>`, `MSS_failure`

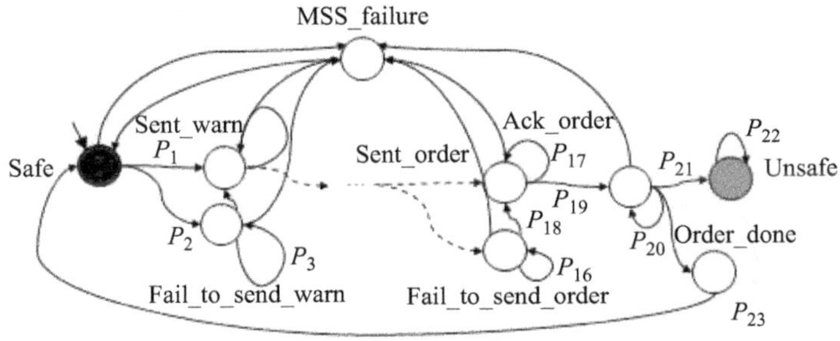

Figure 15.6 A sketch of the MDP modeling of the system-controller for RS1

This system-controller is normally assumed to be in a "safe" state if no initiating event related to a risk scenario occurs. Otherwise, it performs the sequence of RCMs according to the associated risk scenario developed in the previous phases. The "unsafe" end state is reached if no sequence of RCMs succeeds in bringing the system into a safe condition. The `MSS_failure` state allows the encoding of a generic failure which can occur during the execution of any RCM.

In the following paragraph, the list of states defined for the other MDPs is described.

MDP-sensors states: `working` (initial state), `failure` (end state), `send_data`, `fail_to_send_data`
MDP-device states: `working` (initial state), `failure` (end state), `send_ack`, `fail_to_send_ack`, `receive_data`, `fail_to_receive_data`
MDP-patient states: `in_safe_area` (initial state), `unlocated`, `out_of_safe_area`, `aware_of_warning`
MDP-caregiver states: `take_order`, `perf-orming_order`, `order_done`, `order_failed`, `aware_of_alarm`, `aware_of_order`

The states of sensors and devices intuitively follow their possible operations. In particular, the `working` and `failure` states abstract away the detailed internal behavior of such components in order to only model that which is important from the risk analysis perspective.

As for the MDP-patients, their states also capture what is strictly necessary for the RS1. With `in_safe_area` we are able to model patients following the examination procedure within safe areas. When a patient moves through or next to a contaminated area, this situation is modeled as a transition into the state `out_safe_area`. Since the patient is located via an RFID, a case in which the sensor fails to transmit data to the system controller is taken into account (`unlocated` state). The `aware_of_warning` models the patient's mental state of having an awareness of the warning sent by the system controller.

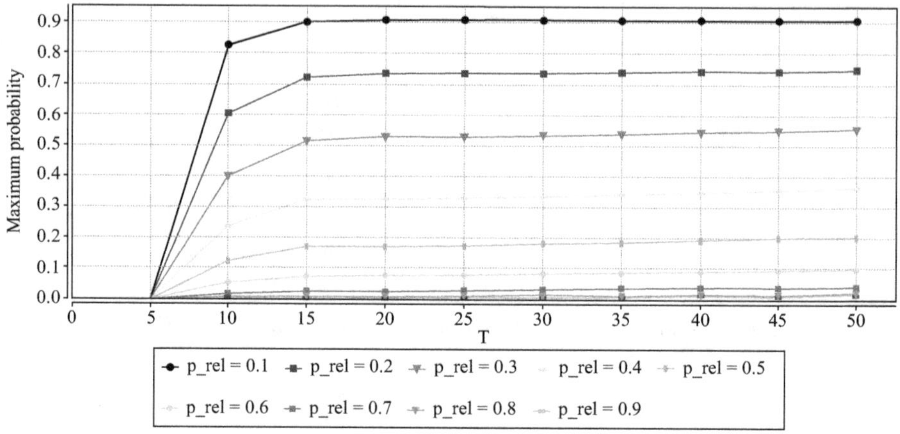

Figure 15.7 Probability of reaching the hazardous situation RS1

The states describing the caregiver category's behavior follow the same logic described for the patient model. The four states listed at the first point capture the caregiver's actions in, respectively, taking up, performing, and successfully/unsuccessfully accomplishing the order. The `aware_of_*` states model the caregiver's awareness, or not, of the received alarm and order.

15.5.5 PRM analysis and risk evaluation

Model checking enables us to conduct two important risk analysis and assessment classes on PRMs. The first is supportive to the system development life cycle, since it gives information on how the probability of incurring a HS (reaching the end state) varies with the change of the transition probabilities within the model. One important analysis we report regards varying the transition probabilities of sensor/device failure states in order to quantify the impact of such a variation on the probability of reaching a successful/unsuccessful end state. During the development life cycle of the system, such an analysis is potentially helpful because designers could fix the maximum probability of an expected HS, and could analyze the best combination of RCMs and quality of the system components necessary to reach the expected goal. The second class allows an evaluation of the trend of the probability of reaching the end state with respect to time. This is the major advantage of using this approach in place of traditional PRA techniques such as ET/FT in that it allows an identification of the time-dependent critical point beyond which the value of probability exceeds the designers' expectations.

The graph in Figure 15.7 represents both classes of analysis conducted on the model by exploiting the probabilistic model checker PRISM. It shows the different probability trends of reaching the unsafe end state of the RS1 with respect to nine different failure probabilities of the devices composing the system. Each probability trend is obtained by iterating the model checking procedure over a diverse time value. In particular, the evaluation is computed by expressing the probability of reaching the unsafe end state within T time units, which in the Probabilistic Computation Tree Logic is formalized as

$$P_{\max=?}\ [\mathrm{F}_{<T}\ \text{`}\mathtt{unsafe\_state}\text{'}] \tag{15.2}$$

By looking at the graph of one specific device probability failure, it is possible to define a threshold point beyond which the variation of risk is negligible with respect to the time. On the other hand, valuable information is obtained by examining the risk level of each graph with respect to the variation of the probability of failure. In particular, it must be noted that the risk changes by ten orders of magnitude when the probability of failure changes from 0.7 to 0.8, whereas this does not happen between 0.8 and 0.9. On the basis of this information and by taking into account also the cost of choosing a device component of a superior quality, the level of risk 0.8 could be selected as a good compromise.

The analysis shows that if a patient moves into an unsafe area and the system has not yet issued an alarm (i.e., the premise of the implication holds), then the probability that the system acts by issuing the alarm within T time units is more than 99%. This property allows us to collect evidence about the safety and effectiveness of the system.

$$\begin{gathered}\mathtt{pat\_in\_unsafe\_area} \wedge \neg\mathtt{alarm\_sent} \implies \\ P_{\max>=0.99}[\mathtt{pat\_in\_unsafe\_area}\ U_{[0,T]}\ \mathtt{alarm\_sent}]\end{gathered} \tag{15.3}$$

Chapter 16
Requirements management

16.1 Background

A requirement of a software system is a characteristic that the system should have, a functionality that it should support, or a constraint that it should hold.

A formal definition was provided by IEEE Std 610.12 [49],

"A **requirement** is (1) a condition or capability needed by the end user to achieve a goal, (2) a condition or capability that is to be passed by a system to satisfy a specification standard or other formally imposed documents, (3) a document representation of a condition or capability as in 1 and 2 above."

For its part, the FDA has also clarified that a requirement is any sort of need or expectation for the system, and it reflects the needs of the customer. Requirements may be contractual, market based, or statutory.

Traditionally, software requirements are stated in functional terms and are defined, refined, and updated as a development project progresses. More recent approaches have enabled the collection of software requirements as Use Cases or User Stories. An accurate and complete documentation of the software requirements is the basis for a successful validation of the resulting software.

A **specification** is defined as "a document that states requirements." There are many different kinds of specifications such as, for example, those for system requirements, software requirements, software design, software testing, software integration, and software maintenance. All these documents are related to "specified requirements" and are the output of design activities for which different kinds of verification may be necessary.

Within the medical industry, one major concern is requirements traceability. Traceability is defined again by IEEE Std 610.12 as "(1) the degree to which a relationship can be established between two or more products of the development process, especially products having a predecessor–successor or master–subordinate relationship to one another; (2) the identification and documentation of derivation paths (upward) and allocation or flowdown paths (downward) of work products in the work product hierarchy; (3) the degree to which each element in a software development product establishes its reason for existing; and (4) discernable association among two or more logical entities, such as requirements, system elements, verifications, or tasks." Requirements traceability, in particular, is defined as "the ability to describe and follow the life of a requirement in both a forward and backward direction (i.e., from its origins, through its development and specification, to its subsequent

Table 16.1 Main causes of software project failures

Problem	%
Incomplete requirements	13.1
Low customer involvement	12.4
Lack of resources	10.6
Unrealistic expectations	9.9
Lack of management support	9.3
Changes in the requirements	8.7
Lack of planning	8.1
Useless requirements	7.5

deployment and use, and through periods of ongoing refinement and iteration in any of these phases)." Requirements traceability is crucial for software verification and validation.

Requirements engineering is the process of finding out, analyzing, prioritizing, documenting, checking, and tracing.

Requirements are the base of all software systems, and their elicitation, analysis, and specification cause a variety of problems independently of the Software Development Life Cycle (SDLF) adopted. In particular, requirements variability is a major challenge for all software projects. According to a study by the Standish Group [43], most of the main factors for software project failure, namely "incomplete requirements," "low customer involvement," "unrealistic expectations," "changes in the requirements," and "useless requirements," are related to requirements, as shown in Table 16.1.

On the other hand, the introduction of faults at the engineering requirements phase may be very costly as shown by the chart in Figure 16.1. Boehm performed one of the first cost studies to determine the cost factors associated with fixing defects. The cost increases up to 100 times when the defect is found late in the operation phase.

16.2 Types of requirements

First, requirements may be grouped according to different levels of abstraction:

Business requirements are the high-level goals/objectives of any of the stakeholders interested in the system. At this layer, business goals are described, which are affected by the organization's vision and scope.

User requirements represent the user need to have or benefit in having the system. Of course, stakeholders and users are generally different people/organizations. This type of requirement is related to what the use that clients will be able to make of the system once available. Users typically have different goals from those of the stakeholders. Additionally, the constraints under which the system must operate are different from those proposed by the stakeholders.

System requirements are detailed descriptions of the system's functions, services, and operational constraints. They are very close to the functionalities that the developers must implement to achieve the desired support from the system.

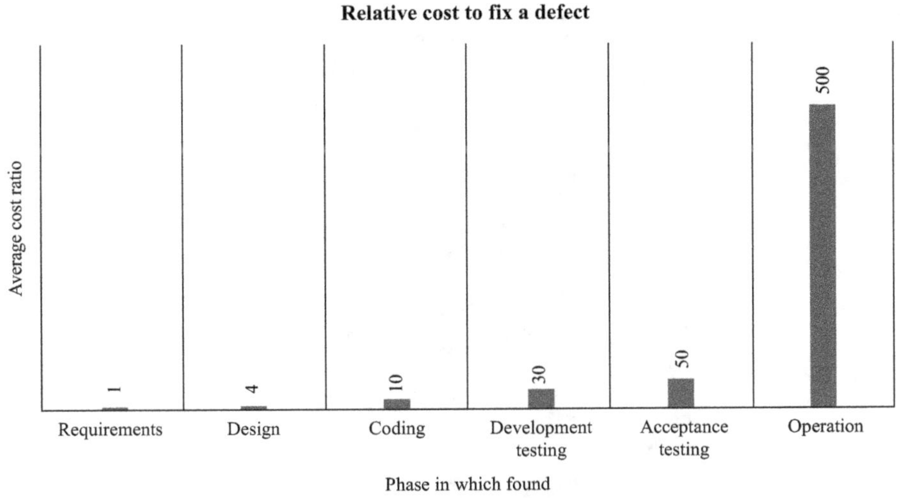

Figure 16.1 Relative cost to fix a defect

Another useful classification of requirements is in terms of functional or nonfunctional requirements.

Functional requirements define services or functions the system must provide. They may also describe how the system reacts to certain inputs. Functional requirements may be directly mapped into software functionalities that the developers must integrate within the system to enable users to accomplish their tasks.

Nonfunctional requirements are constraints on the services or functionalities provided by the system. They restrict how the system must function and often apply to the system as a whole rather than to individual functions or components. Nonfunctional requirements define properties and constraints related to characteristics such as security, privacy, dependability, performance, and underlying operating systems (Figure 16.2).

There exist several quality characteristics for requirements:

1. Correct—They are correct when they reflect the actual needs of the stakeholders;
2. Complete—They should include a complete description of all the characteristics and facilities required;
3. Unambiguous—They should be clear and not lead to misinterpretations or different interpretations by different people;
4. Consistent—There should be no contradictions or conflicts in the descriptions of the system facilities;
5. Verifiable—The requirements should be written in such a way that the final system can be tested against all of them, and it must be possible to determine whether or not the system meets the requirements.

Table 16.2 reports some requirements for the system for the Nuclear Medicine Department case study introduced in Chapter 11.

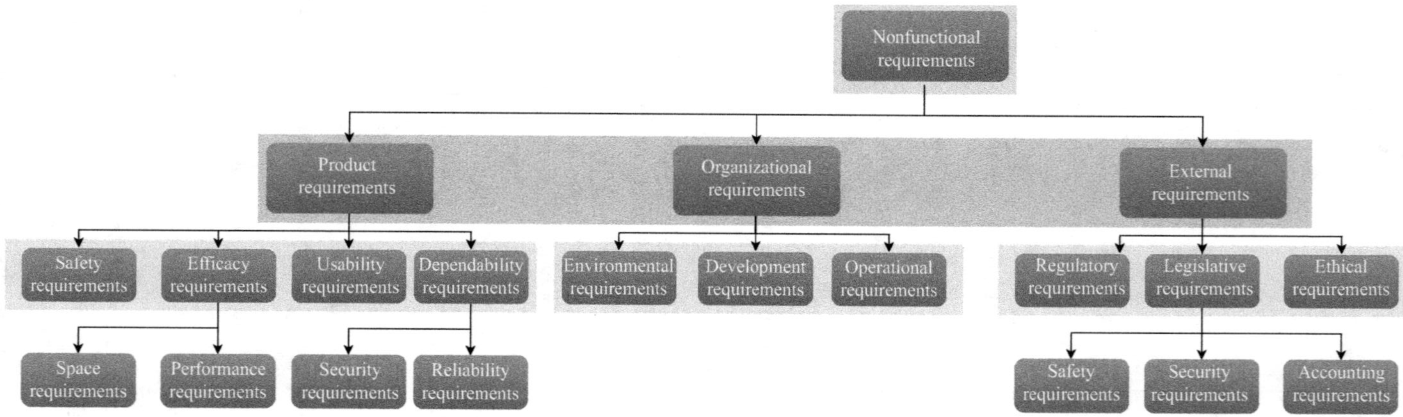

Figure 16.2 Nonfunctional requirements

Table 16.2 Example of requirements for the Nuclear Medicine Department case study

		Functional		Nonfunctional
Business requirements	**BR1.**	The system must enable a reduction in the exposure of the personnel to radiation		
User requirements	**UR1.**	The system must allow the doctor to monitor the level of radiation of a patient		
	UR2.	The system must allow the doctor to call a patient to the *InjectionRoom*		
	UR3.	The system must allow the patient to locate which room she/he has to move to		
System requirements	**SR1.**	The system must advise the patient when she/he tries to move into a forbidden area	**SR2.**	The system must be able to handle at least five patients contemporaneously
			SR3.	Any location error must be of less than 1 m
			SR4.	System interfaces must be able to resize depending on the user's device (e.g., SmartPhone or Tablet)

It is worth noting that at the business requirements layer the stakeholder is the nuclear department top management who requires the development of the system, whereas at the user requirements layer the users are department staff and patients. Therefore, a major business goal is to "Reduce the exposure of the personnel to radiation," whereas some services that the system must offer to its users are "Locate the room where to move," "Locate the patient," "Monitor the level of radiation of a patient," and "Call a patient to the *InjectionRoom*." Examples of system requirements are "The system must advise the patient when she/he tries to move into a forbidden area" (functional), "The system must be able to handle at least five patients contemporaneously" (performance), "Any location error must be of less than 1 m" (performance), and "System interfaces must be able to resize depending on the user's device (e.g., SmartPhone or Tablet)" (usability).

16.3 Requirements development

Requirements development comprises three main activities, namely requirements elicitation (gathering), requirements specification, and requirements validation.

16.3.1 Requirements elicitation

The terms gathering or eliciting requirements are used to refer to the process of finding out what the needs, desired services, and constraints are for a given software system. Gathering and eliciting are often used indifferently, although they refer to two distinct practices. Requirements gathering is the term mostly used in traditional Software Development Life Cycles and deals with the collection of observable requirements from the domain, the analysis of the documentation, and any interviews with the stakeholders. Elicitation is more commonly adopted in Agile environments and places more emphasis on the need to discover and bring out hidden nondirectly observable requirements. It is the effort to extract information from stakeholders and experts of the application domain. It is a set of techniques that may be applied, appropriately, during the requirements phase.

Some of the techniques for eliciting requirements are:

Brainstorming, which can produce numerous ideas in a limited time. This technique may establish ground rules and criteria for the ranking of the ideas.

Interviews, a practice which involves stakeholders and experts of the application domain and represents the most rapid way of acquiring information.

Document analysis, which extracts "knowledge" from existing documentation and is one of the main sources of information when experts are not available.

Process analysis, which is the way the requirement analyst can understand the kind of work, and activities are performed within the organization for which the system will be realized.

Interface analysis, which allows the analyst to extract requirements from the analysis of other operating systems (e.g., applications that will be replaced by the system under construction).

Prototyping, which may be extremely useful to clarify unclear requirements.

16.3.2 Requirements specification

As the list of requirements is available, they must be specified in a more formal format. Traditional approaches require that requirements are specified in a Software Requirements Specification (SRS) document. Figure 17.5 shows the contents of the IEEE SRS Standard document [50] (Figure 16.3).

The characteristics of a good SRS are: correct, unambiguous, verifiable, traceable, consistent, complete, modifiable, and ranked for importance and/or stability [51]. The first four of those characteristics hold if, and only if, they hold for every requirement reported in the SRS.

Moreover, an SRS is complete if it includes all significant requirements, whether relating to functionality, design constraints, performance, or external interfaces and if it defines the responses of the system to all possible classes of input data taking into account both valid and invalid values.

The SRS is modifiable if its structure and style are such that any necessary changes to the requirements can be made easily, completely, and consistently.

Finally, an SRS is ranked for importance and/or stability if each requirement has an identifier to indicate either the importance or stability of that particular requirement.

Figure 16.3 IEEE Std 820—Software Requirements Specification table of contents

It is worth noting that the system features in the SRS document may be functional requirements, user classes, classes/objects, stimulus–responses, or information flaws. Another very common way of describing a system's requirements is by means of Use Cases. A Use Case is a sort of contract between the stakeholders/users and the system to achieve a specific goal. A Use Case represents a point of access to the system for its users and describes a service or functionality that is activated by the user. Use Case modeling enables a specification of the behavior of a reactive system: the user (called the Actor) stimulates the system by executing one of its Use Cases; next, an interaction between the system and the user is developed until the completion of the Use Case, which ends with the achievement of the goal, in the case of a "Main success scenario," or another state as described by an alternative scenario called an "Extension."

Figure 16.4 shows part of the Use Case diagram for the case study. The doctor can either log in or monitor the radiation level of a specific patient. As specified by the "«include»" relation, the latter Use Case requires that the patient be logged in before proceeding. Table 16.3, instead, reports further details about the execution of such Use Cases. Indeed, some execution scenarios are specified. With respect to the log-in Use Case, the main success scenario describes the interactions developed between the user and the system in the case that the user's objective (to be logged in)

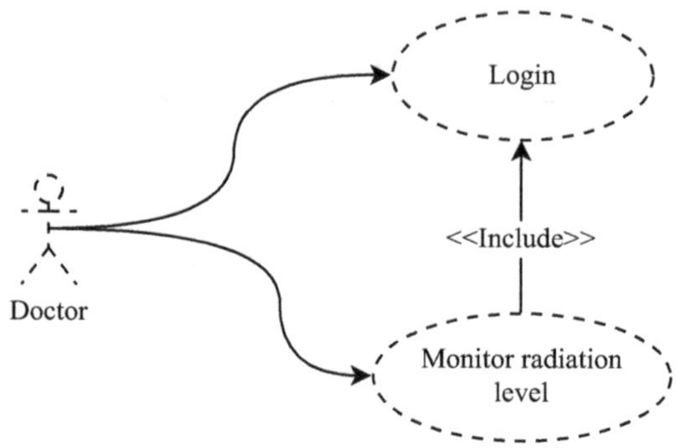

Figure 16.4 An example of a use case diagram

Table 16.3 Use case scenarios

Use case	Scenarios
UC1: Log-in	*Main success scenario*: UCS1.1 1. The doctor needs to log in 2. The system needs a username and password 3. The doctor provides her/his username and password 4. The system checks for the username and password and allows the doctor to log in *Extension*: UCS1.2 4.a The system checks for the username and password and requires the user to retype the username and password, or to leave 5.a The doctor provides a new username and password 6.a The system checks for the username and password. If correct, it allows the doctor to log in. Otherwise, it returns to 4.a
UC2: Monitor radiation level	*Main success scenario*: UCS2.1 1. The doctor requires to monitor the radiation level 2. The system requires to select the patient 3. The doctor selects the patient 4. The system provides the radiation level of the selected patient

is definitively achieved. The extension, instead, specifies an alternative sequence of interactions in the case that it is not possible to achieve the user's objective.

It is worth noting that the interactions described in the scenarios affect the number and content of system interfaces crossed to achieve the user's objective.

In an Agile environment, requirements are documented by means of User Stories as already described in Section 13.3.2.

A User Story is just something that a user wants. They may be structured in the form "As a…," "I want…," "So that…." Typically, in the Scrum framework, User Stories, along with Epics, feed the Product Backlog for the production.

An Epic is a collection of macro-functions. In other words, it is a "sufficiently" large User Story. Therefore, "Epic" is a label that applies to a large story, which could be broken down into stories that would probably be small enough to be implemented directly.

User Stories may also be categorized into so-called Themes.

16.3.3 Requirements verification and validation

All techniques presented in Part IV are useful for requirements verification and validation. In the following section, we will briefly summarize some of these techniques and add others specifically devised for requirements checking.

- Simple checks—These techniques verify that the requirements are well written, for example, according to the SRS document standard guide or quality criteria established by a quality policy. Other kinds of simple checks are those performed by using traceability techniques such as checking that all elicitation notes are covered, tracing between different levels of requirements, and verifying goals against tasks, features, and requirements;
- Reviews and inspections—A team analyzes the requirements, looks for potential issues, invites participants to meetings to discuss any problems, and identifies actions to address those problems;
- Prototyping—This is very effective for requirements validation against user expectations. It demonstrates the requirements and helps stakeholders discover problems. Different types of prototypes can be realized, from paper sketches of the system interfaces to formal executable models/specifications passing through the running and testing of the software components;
- Functional test design—Functional tests at the system level, which should be derived from the requirements specification, may reveal defects in the specification even before the design and building of the system. Indeed, as an example, missing or ambiguous information in the requirements description may make it difficult to formulate tests;
- Model checking—This is a formal model-based verification and validation. It uses automatized tools (e.g., model checkers) to check the model against certain formal properties (e.g., temporal logic constraints).

16.4 Requirements traceability

Requirements traceability is substantially related to three aspects, namely analyzing: (1) the life of a requirement, from elicitation to implementation; (2) how requirements impact on each other, and how they impact on other artifacts (e.g., design models, tests, and code) and vice versa; and (3) how requirements may be decomposed.

Traceability deals with establishing and maintaining relationships (among requirements or between requirements and other artifacts), which may be of different types:

- Satisfaction: A system requirement, or a group of system requirements, satisfies a user requirement;

Table 16.4 Example of traceability matrix

User requirement	Use case	Use case scenario	Test case
UR1	UC2	UCS2.1	TCx

- Verification: A test case verifies a requirement;
- Dependency: A requirement depends on one or more other requirements.

Basic traceability establishes a link between one or more elements. It is possible to specify the type of traceability, which relates the relationship type with its associated semantics. In addition, it is possible to specify the so-called rich traceability, which includes additional semantic information, for example, the rationale explaining why a system requirement satisfies a specific user requirement.

The rich traceability approach is particularly valuable in heavily regulated industries and safety-critical systems where audit trails of decisions made are vitally important to provide assurance and reduce risks.

Once traceability relationships have been identified, they can be represented and reported on by means of traceability matrics. The extract of a traceability matrix reported in Table 16.4 shows that the user requirement UR1 of the case study (see Table 16.2) is implemented by the Use Case UC2 (see Table 16.3), and its main success scenario (UCS2.1) is verified by test case TCx (not developed in this section).

Requirements traceability was originally used to demonstrate that contractual requirements had been addressed, but there are other relevant reasons why requirements traceability provides value:

- Quality audit: Traceability helps people to navigate the project and see why particular requirements, designs, tests, etc. exist. This is particularly useful during a quality audit because it enables the project manager to demonstrate that the regulation's requirements have been met.
- Coverage: Traceability enables an analysis and demonstration of the number of user requirements covered.
- Impact analysis: Traceability helps the development team to understand what kind of impact a requirement change may have on system requirements, test cases, and software components. Conversely, it is possible to understand the system requirements and user requirements affected by a change in the software (e.g., an API upgrade).

Chapter 17
Design controls and development management

17.1 Background

Design controls were introduced in the good manufacturing practice requirements and become effective on June 1, 1997. They are now part of the 21 CFR 820 Quality System Regulations. Specifically, Part 30 of the regulation specifies the requirements that affect the development of all Medical Devices falling into Classes IIa and IIb, and some in Class I, listed in the regulation itself.

Design controls are management practices such as policies, processes, and procedures that apply to control design activities. The goal is to control the design process in order to ensure that the system specifications meet the user needs and intended use.

Design controls are necessary due to the fact that the most frequent causes of recalls are related to design and software. Indeed, 90% of all software-related-device failures from 1983 to 1989 were due to design-related errors as stated in the "FDA Medical Device Regulation from Premarket Review to Recall" study. Moreover, 44% of voluntary recalls from October 1983 to September 1989 may have been prevented by adequate design controls, according to "Device Recalls: A Study of Quality Problems" and "QS Regulations (Final Rule)."

Design controls must be set up for a new product after the completion of the project-initiating phase and before the project-planning phase; that is, once it has been determined that the product is feasible and the decision to start the project has been taken.

The real essence of design controls is to provide evidence that the system has been designed to be safe. ISO 13485 has the same intention, discussed in Section 7.3—design and development, although the terminology is slightly different. Table 17.1 compares the ISO 13485 clauses regarding design and development with the FDA clauses for design controls.

Both regulations require that the producer maintains documentation and records throughout the product development process. In the case of a system that must be compliant with the FDA regulation, the design history file (DHF) is the place to annotate all the design controls evidence. ISO 13485, instead, does not mention explicitly a document for the registration of such evidence. However, it is expected that the producer maintains records of design and development activities.

Table 17.1 FDA 820.30 design controls versus ISO 13485 design and development clause

FDA 820.30 design controls	ISO 13485 design and development clause
1. General	7.3.1 Design
2. Design and development planning	7.3.2 Design and development planning
3. Design input	7.3.3 Design and development inputs
4. Design output	7.3.4 Design and development outputs
5. Design review	7.3.5 Design and development review
6. Design verification	7.3.6 Design and development verification
7. Design validation	7.3.7 Design and development validation
8. Design transfer	7.3.8 Design and development transfer
9. Design changes	7.3.9 Control of design and development changes
10. Design history file	7.3.10 Design and development files

A best practice is to realize traceability matrixes to show the relationships between user needs, user requirements, system requirements, design artifacts, test cases, and software components.

17.2 Design controls

The list of FDA design controls has already been reported in Section 3.2 as defined by FDA 21CFR 820.30. In the following section, we provide some comments:

1. The design and development plan—This establishes the overall development plan, describes the development process, and defines the design activities and responsibilities. It clarifies the roles of all participants in the development process. The plan also outlines the medical purpose (intended use), which affects the software classification and consequently the quality requirements (regulation clauses) that apply.
2. Design inputs—These deal with the physical and performance characteristics of the system, which are used as a foundation for the design. They establish requirements that will ensure that the system will meet the needs of the intended users. They may be collected in a requirements document (e.g., a software requirements specification).
3. Design outputs—They represent the result of a design activity performed at any design phase of the development process. Design outputs are the system model that must meet the design input requirements.
4. Design review—This is a planned, formal, and documented review of a design result performed to ensure that the design outputs are meeting the design input requirements. The results of any review meeting are recorded in the DHF (or in any other document for an ISO 13485 QMS). A formal review (a minimum of two reviews is necessary) requires the participation of representatives of all functions involved in the design stage under review, an individual or group of people who do not have any direct responsibility for the design stage under review, and any necessary specialist.

5. Design verification—This assesses the conformance of the design output to the requirements and confirms and documents that it meets the design input requirements. Substantially, it verifies that the product is correct.
6. Design validation—This assesses, under defined operating conditions, that the product conforms to the user needs and intended use. The design validation shall include a software validation and risk analysis.
7. Design transfer—This ensures that the design specification of the system is correctly translated into specifications for the production.
8. Design changes—These deal with the changes that any design artifact may experience during the system life cycle. Such changes are to be documented, verified, and validated. They are also to be reviewed and approved again before their implementation.
9. DHF—A DHF should be established and maintained for each system. It should contain or reference the records necessary to demonstrate that the design was developed in accordance with the approved design plan. It is a collection of registrations reporting on all the outputs obtained from the previous controls.

Design controls may be applied to any development process model. In the case of a Waterfall model, the FDA Design Control Guidance For Medical Device Manufacturers [52] summarizes the influence of design controls on the design process by means of the model shown in Figure 17.1.

The workflow starts with an analysis of the user needs; next, the system is designed and, successively, its design is evaluated and transferred to production, and the system is realized.

The design process may be iterative or subdivided into subphases. In this latter case, at each stage of the design process, once the design input has been reviewed and its requirements are determined to be acceptable, such requirements have been transformed into specifications, which are the design output. Next, it is verified that this design output conforms to the design input requirements and then this becomes the new design input for the next phase of the design process. In this way, the design input requirements are converted into a design that conforms to those requirements.

The design reviews may be conducted at different points in the design process. As an example, a review of the design input is performed before starting the design activity in order to ensure that the design input requirements are correct, complete, unambiguous, consistent, and verifiable. Other controls may be conducted to assure that the design is adequate prior to realizing the prototype. Generally, reviews are used to ensure that the result of an activity achieves an acceptable level of quality.

Design validation includes verification and aims at establishing whether or not the device/system actually satisfies the intended use and user needs.

17.3 Design control and development templates

The design control templates reported in the following section represent a suite of documents that have been established for the design and development subsystems of the eHealthPlatform case study falling within Classes I, IIa, or IIb. No Class III

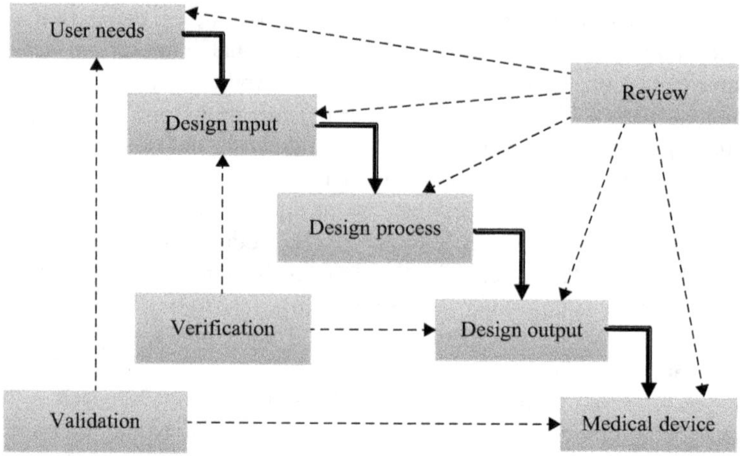

Figure 17.1 FDA design controls

Table 17.2 Documents established for the eHealthPlatform case study

Document	**Class I**	**Class IIa**	**Class IIb**
Intended use	✓	✓	✓
Risk management file	✓	✓	✓
Software development plan	✓	✓	✓
Software requirements specification	✓	✓	✓
Software architectural design		✓	✓
Software detailed design		High level	Low level
Test plan	✓	✓	✓
Test case specification		✓	✓
Test procedure specification		✓	✓
Test incident report		✓	✓
Test summary report	✓	✓	✓
Review report		✓	✓
Meeting report		✓	✓

subsystems were present. The list of documents required for any class of subsystem is reported in Table 17.2.

17.3.1 Intended use template

The content of the intended use template document, shown in Figure 17.2, is

1. *Objectives of the document*—This section is intended to clarify the objective, intended use, and intended users of the system under development;
2. *Description of the software platform*
   (i) *General description*—This section provides a general description of the system under development;

2. OBJECTIVES OF THE DOCUMENT 4

3. DESCRIPTION OF THE SOFTWARE PLATFORM 5

3.1. General description 5
3.1.1. How to use 5
3.1.2. Interactions with other Medical Devices 5
3.1.3. Intended Users 5
3.1.4. Interactions between the software platform and intended users 5
3.1.5. Intended use environment 5

***3.2. Medical Purpose* 6**

***3.3. Indications* 6**
3.3.1. Description of pathologies involved 6
3.3.2. Description of the patients involved 6
3.3.3. Counter-indications and warnings 6

4. QUALIFICATION AND CLASSIFICATION 7

Figure 17.2 Content of the intended use template document

(a) *How to use*—This section describes briefly how to use the technologies that will be developed;
(b) *Interactions with other medical devices*—This section clarifies the interactions (if any) between the system under construction and other existing medical devices. The way these systems may interact (e.g., launching services, exchanging messages, etc.) and the class of risk for these devices must be specified;
(c) *Intended users*—This section reports the list of final users of the system under construction;
(d) *Interactions between the software platform and intended users*—This section describes the way any intended users may interact with the system. It is also important to provide information about the duration of the interactions;
(e) *Intended use environment*—This section clarifies the environment in which the system under construction will be used (e.g., operating room, home, etc.);

(ii) *Medical purpose*—This section provides information about the intended use of the technology under construction, whether it is for diagnosis, treatment, or any other reason for its qualification as a medical device and its classification with respect to the categories of risk of the European Union Directive 93/42/EEC should also be reported;
(iii) *Indications*—This section provides any prescription and indication on the use of the system;
(a) *Description of pathologies involved*—This section reports the list of pathologies that may benefit (diagnosed, treated, or involved) from the use of the system under construction;

   (b) *Description of the patients involved*—This section specifies the classes of patients involved by the system under construction (e.g., a system for tele-monitoring EEG signals includes patients with cardiac problems among its primary users);
   (c) *Counterindications and warnings*—This section declares any counterindications and warnings (if any) for the use of the technology;
3. *Qualification and classification*—This section reports on the qualification and classification process.

17.3.2 Risk management file template

The content of the risk management file template document, shown in Figure 17.3, is

1. *Description of the document*—This section presents the main objectives of the document and shows the references to other project documents, standards, and regulations of interest;
2. *Risk management*—This section presents the risk management approach and process, specifies the responsibilities assumed by people for the risk management, reports the skills required by people involved in the risk management process, presents useful information concerning the production and maintenance of the software in accordance with the organization's policies and strategies, and specifies the criteria adopted to evaluate and accept the risks;
3. *Risk analysis*—This section deals with risk analysis. It identifies risks related to any possible improper use of the technologies, to human factors, or to system failures. Next, the assessment of such risks, as well as the risk control measures, residual risks, and emerging risks, is documented. Finally, a traceability matrix reports the relationships between the risks and the software requirements.

17.3.3 Software development plan template

The content of the software development plan template document, shown in Figure 17.4, is

1. *Software development life cycle*—This section describes the Software Development Life Cycle (SDLC) adopted for the development of the system under construction. It lists and describes all the tools that will be adopted at any stage of the SDLC, including Requirements management, Software Design, Programming, Testing and Incident Management, and Configuration management. Moreover, the section refers to or describes other processes such as, for example, risk management and testing, and establishes rules and standards adopted for the development of the software system;
2. *Participants and responsibilities* —This section lists all the participants (teams or organizations) in the development and establishes roles and responsibilities.

Figure 17.3 Content of the risk management file template document

17.3.4 Software requirements specification template

The content of the software requirements specification template document, shown in Figure 17.5, is

1. *Objectives*—This section reports the main objectives of the document;
2. *Business scenario*—This section describes the business scenario. The description includes a scenario overview, a definition of business terms of common use in

Figure 17.4 Content of the software development plan template document

the scenario, an identification of the stakeholders, and a specification of all the business processes;

3. *Use cases*—This section specifies the system's use cases. In particular, for each use case, the name, description and priority, main success scenario, extensions (if any), and functional requirements are reported;
4. *Nonfunctional requirements*—This section specifies all nonfunctional requirements such as performance, security, privacy, etc.;
5. *Usability requirements*—This section focuses on usability requirements with particular interest in the user interface requirements and human factors.

17.3.5 Software architectural design template

The content of the software architectural design template document, shown in Figure 17.6, is

1. *Objectives* —This section reports the main objectives of the document;
2. *Resources*—This section specifies the hardware and software resources needed by the software under construction (i.e., its logical components) to implement the software requirements;
3. *Architecture*—This section describes the logical architecture for the system under construction. In particular, it provides a logical view of the components and packages. For each component, the description and scope, list of services offered, data model, and interfaces are provided. A relevant focus is on

Figure 17.5 Content of the software requirements specification template document

the Software of Unknown Provenance (SOUP) and Commercial Off-the-Shelf (COTS) components;

4. *Traceability software requirements/components*—This section specifies the relationships between the software requirements and logical components.

17.3.6 Software detailed design template

The content of the software detailed design template document, shown in Figure 17.7, is

1. *Objectives*—This section reports the main objectives of the document;

Figure 17.6 Content of the software architectural design template document

2. *System object model*—This section specifies the components architecture in terms of objects and classes. It reports static models for such entities;
3. *Dynamic models*—This section specifies the dynamic behavior of the system by means of models such as the collaboration, activity, and state models;
4. *Traceability software requirements/objects*—This section specifies the relationships between the software requirements and objects.

17.3.7 Test plan template

The content of the test plan template document, shown in Figure 17.8, is

1. *Objectives*—This section reports the main objectives of the document;
2. *Test environment*—This section specifies the test environment. In particular, it describes the test team, responsibilities, hardware and software characteristics of the test environment, testing tools adopted, and requirements and constraints for the installation, configuration and maintenance of the testing platform;
3. *Test management*—This section describes the testing process, timetables, testing categories (e.g., usability testing, functional testing, etc.), techniques and strategies for testing (e.g., random testing, risk-based testing, statement coverage, equivalence partitioning, etc.), and acceptance criteria;

Figure 17.7 Content of the software detailed design template document

4. *Test planning*—This section presents the current test planning. In particular, for each test suite, it reports the version, objectives, components to be tested, and list of test cases.

17.3.8 Test case specification template

The test plan template document comprises the following information for each test case of a suite:

1. *Test case identifier*—This is an identification code for the test case;
2. *Test item*—This clarifies the software item/items tested and its/their version;
3. *Test precondition*—This is the precondition (initial state) to ensure prior to the execution of the test;
4. *Test input specification*—This is the set of input data for the test;
5. *Test output specification*—This is the expected result of the test;
6. *Test case dependencies*—This specifies the test case dependencies (if any) from other test cases;
7. *Procedural requirements*—This specifies the procedural requirements (if any);
8. *Traceability*—This specifies the requirements tested.

17.3.9 Test procedure specification template

The test procedure specification template document comprises the following information for each test procedure:

1. *Test procedure identifier*—This is an identification code for the test procedure;
2. *Purpose*—This describes the procedure and lists all the test cases involved;

Figure 17.8 Content of the test plan template document

3. *Requirements*—This provides the requirements for executing the test (e.g., manual versus automatic testing, test environment, etc.);
4. *Procedure steps*—This is the list of steps to accomplish for the procedure.

17.3.10 Test incident report template

The test incident report template document comprises the following information for each anomaly:

1. *Anomaly identifier*—This is an identification code for the anomaly detected;
2. *Test case*—This is the identifier of the test case that has failed;
3. *Description*—This is a description of the deviation observed from the expected behavior of the system. This section also includes information on the components version, test environment, test tools, etc.

Figure 17.9 Content of the test summary report template document

17.3.11 Test summary report template

The content of the test summary report template document, shown in Figure 17.9, is

1. *Objectives*—This section reports the main objectives of the document;
2. *Test results*—This section illustrates for each test suit the results of the testing. In particular, it provides information about the test environment, and software version, and a list of tests passed and failed;
3. *Test planning*—This section reports conclusions and recommendations.

17.3.12 Review report template

The review report template document comprises the following information:

1. *Review identification number*—This is an identification number for the review;
2. *Reviewers*—This is the list of reviewers;
3. *Artifact*—This is the artifact reviewed;
4. *Defects and anomalies*—This is the list of defects and anomalies identified, along with a detailed description;
5. *Countermeasures*—This is the list of the countermeasures approved in response to the anomalies and defects found;
6. *Concluding remarks*—These are the concluding remarks.

17.3.13 Meeting report template

The meeting report template document comprises the following information:

1. *Date, place, and arguments discussed*;
2. *List of participants*;
3. *Decisions and actions established to be performed*;
4. *Criticality*;
5. *Concluding remarks*.

Chapter 18

Test management and defect management

18.1 Software testing principles

Before presenting some software testing management strategies and methods, it is worth introducing the seven fundamental principles of software testing, which have been established over several years of experience. These principles can be used as a guideline for both software testing and coding activities.

The following section introduces and describes such principles:

Principle 1—Testing shows the presence of defects. Software testing cannot prove the full correctness of a system. On the contrary, it can show that some defects are present. Even if a testing campaign is performed using a certain test suite and no failures are observed, this is not a proof of full correctness. Indeed, as long as a software campaign is unable to exercise a system with respect to all possible combinations of input data and internal conditions (Principle 2), there will always be the possibility of deviation from the expected behavior. Consequently, nobody can ensure that the system will not fail against certain untested conditions or when executing certain untested parts of the code.

This principle influences the approach that a software team should adopt when undertaking its work and designing test cases. Indeed, the goal should be that of making the system fail and, consequently, of letting defects emerge, instead of trying to prove that the system is correct.

Principle 2—Exhaustive testing is impossible. Exhaustive software testing, which would require exercising the system against all possible combinations of input data and internal conditions, is not feasible except for trivial cases. Indeed, if we consider a simple application that has only one internal state and eight input parameters, each of which can assume six distinct values, the number of possible combinations is 8^6=262,144. If we suppose that we need a mean of 10 min to design, write and execute a test case, and to compare the observed result with that expected (10 min may be sufficient for the first test cases, but the time necessary to design a new test case increases with the number of test cases that we have already realized), exhaustive testing would take almost 5 years.

Instead of exhaustive testing, other testing strategies are adopted including a risk-based one, which is useful in order to focus testing efforts.

Principle 3—Early testing. Testing activities should start as early as possible. In this way, it is possible to reduce development and testing costs. Indeed, the cost of fixing increases up to 10 times when a defect of a requirement is detected during a dynamic testing activity and up to 100 times when it is observed after the deployment to the client, compared to discovering it during the requirements elicitation phase. For this reason, static testing is definitively recommended as a means of reducing both the number of defects and the costs of fixing.

Principle 4—Defect clustering. The distribution of defects in a software system is not uniform among its parts. The defect density depends, among other elements, on the complexity of the part of the software under analysis. The greater the complexity, the greater the probability and density of defects.

As a consequence, the testing effort should not be uniform, but it should be proportional to the expected defect density of the part of the software under testing. The Pareto principle, also known as the 80/20 rule, states that 80% of the defects are located in 20% of the software code! This information could be used while testing because if we find one defect in a particular area of the software, there is a pretty high probability of finding others in the same area.

Principle 5—Pesticide paradox. The "Pesticide Paradox" refers to the situation in which the same kinds of test are repeated over and over again. This testing strategy will not be effective as long as different tests exercise the same set of test conditions. To avoid such a paradox, it is necessary that every new test case is designed and executed to verify at least one new (uncovered so far) test condition.

This principle, however, does not apply when a new version or a simple modification of the software is realized. In such a case, indeed, a regression testing is necessary to verify that what "worked" before the modification, still works afterwords.

Principle 6—Testing is context dependent. Testing must be performed differently in different contexts. As an example, a safety-critical medical software system is tested differently from an *e*-commerce portal. The former presents more critical functionalities, and may require to be verified in its final environment (e.g., to exclude intersystem interferences), etc.

Principle 7—Absence-of-errors fallacy. This principle emphasizes the fact that if no failures are observed during a testing campaign, the conclusion cannot be that "everything works fine." Indeed, the most relevant issue is whether or not the software system meets the user's expectations or results in a useless product. Another potential issue to consider is that the absence of failure could be due to an incorrect test case design.

18.2 Software testing strategies

In any case in which the system under realization is nontrivial and exhaustive testing is impracticable, it is necessary to have strategies that allow a maximization of the level of quality achieved by means of testing activities, even if the testing effort is to be concentrated only on a reduced set of test conditions or on a few parts of the software.

The following testing strategies, which are not mutually exclusive, may be chosen to prioritize the tests and allocate the effort; so that a test manager can organize a process to select a finite set of test conditions. Such a process may also help to determine the appropriate effort and resources to provide in each test condition.

Risk-based testing. In risk-based testing, the risks assessed in the risk management process are used to guide the testing activities. In particular, the effort for each risk factor is allocated proportionally to the level of risk. Testers select the test techniques to apply depending on the rigor and level of coverage required by the level of risk. The most important risks are addressed before, and in greater detail than, the less important risks. Testing is a risk mitigation measure and residual risks are assessed after the testing campaign.

Model-based testing. Model-based testing deals with the design and realization of test cases from models of the system and its behavior, such as control flow, data flow, dependency graphs, state transition machines, decision tables, etc. The most relevant benefit of such a strategy is the possibility of automatizing the test case generation. Specific tools are able to process models of the systems—especially finite-state models—and to generate test cases. A cover policy is necessary to select the test conditions to exercise. Several coverage techniques (e.g., Statement, Branch/Decision, and Condition, etc.) have been presented in Chapter 10.

Exploratory testing and **error guessing**. Exploratory testing and error guessing are relatively unstructured approaches to testing. Exploratory testing concerns exploring and acquiring knowledge about the software, how it works, what works, and what does not work. This strategy is very useful when there are limited or no specifications of the system. Testers are involved in minimum planning and maximum test execution, and constantly decide what to test next.

Error guessing is related to the ability of the tester to "guess" and foresee which kind of conditions may be critical and cause the system to fail. Testers use their intuition and rely on their experience of testing similar or related systems, their knowledge of typical design and development errors, and their knowledge of the results of any earlier testing phases.

Random testing. Random testing relies on the random generation of test cases. This may be helpful to complement other testing strategies because of certain properties, such as the absence of human bias, and the simplicity and speed of execution.

18.3 A software testing process

The test process described in the following section, like all other processes, can be viewed with different levels of detail. Each activity can be detailed as a process in itself. The test process here presented is derived from that defined in the International Software Testing Qualifications Board (ISTQB) syllabus and consists of several activities (or processes), which are not performed in a strict sequential order.

As shown in Figure 18.1, Monitoring and Control are umbrella activities that extend throughout the test cycle.

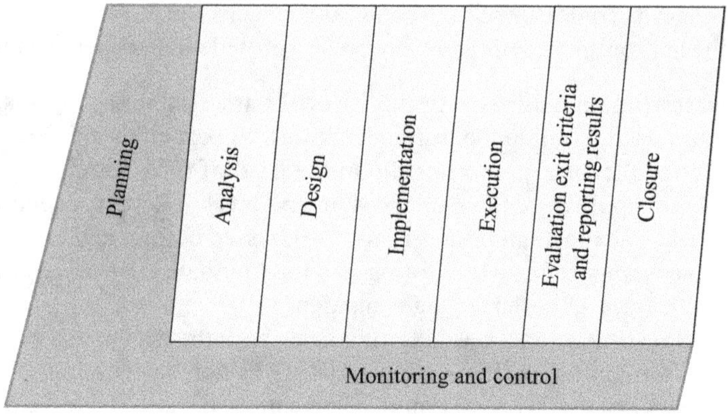

Figure 18.1 A test process

18.3.1 Test planning, monitoring, and control

The processes of planning, monitoring, and controlling the testing are management tasks.

Test planning is the first task, which, however, continues throughout the test cycle and for every iteration. This task concerns the achievement of the test objectives. In particular, the means of achieving the objective is determined during this step, which means identifying the activities and resources necessary to accomplish the mission and the objectives, and continuously updating the plan as other test activities are performed and new results are obtained.

Measure the degree of achievement of the test objectives is part of this step. This requires that metrics are defined during the planning activity.

During the test planning, the organization's test strategies are confirmed or, if necessary, amended. The choice of test strategy influences other results of this activity. For example, if the risk-based testing strategy is adopted, risk analysis guides the test planning with the aim of reducing the identified product risks. Indeed, risk information may be used to determine and allocate a sufficient amount of effort in order to develop tests aiming at mitigating unacceptable risks. Risk assessment also drives the test case prioritization. Likewise, if it is recognized that the design activity will probably be error-prone, the test planning should determine an additional static testing of the design artifacts.

As the testing activities progress, it is necessary to monitor their trend and, in any case in which a relevant deviation from the plans is observed, to undertake control measures.

Checkpoints, in which metrics are used to measure the progress of the testing, are defined during the planning to monitor periodically the state of execution of the project. Possible metrics for this aim are, for example, the percentage of statement coverage, and decision coverage, etc. However, instead of a pure analytical report focused on the values of the previous metrics, a more detailed documentation with a description of the risks mitigated, use cases employed and business requirements verified, etc., is preferred because it better meets the requirements of other stakeholders.

18.3.2 Test analysis

Test analysis defines the test conditions, namely "what" is to be tested. Tests analysts identify the test conditions by analyzing the test basis; i.e., all the documents useful to infer the requirements of the system to be tested. In particular, use cases (in the case of use case driven requirements engineering) and user stories (in the case of agile methods) are part of the test basis.

Test conditions may be defined at different levels of detail. For example, data values or variables may or may not be defined. The level of detail of the test conditions depends on several factors such as the level of detail of the test basis and its quality, the level of testing, the skills of the test analysts and designers, the product risks, the system complexity, etc.

A detailed specification may result in a large number of test conditions. For example, considering the use case UC1 (log-in) in Table 16.3, a high-level test condition could be "Test the main success scenario"; whereas, a few low-level test conditions, for the same use case, could be "T1: Test the log-in operation with a valid username and password"; "T2: Test the log-in operation with an invalid username"; "T3: Test the log-in operation with a not well-formed password"; etc.

The specification of detailed test conditions has some advantages, such as, for example, easing the work test analysis, providing more detailed information for monitoring and control, and providing a simpler and more direct traceability within the testing activities. However, it has also some disadvantages such as the greater amount of effort required and the more difficult maintainability, Nevertheless, specifying the test conditions at a detailed level can be particularly effective when the project is large or complex, or when only a few or no formal requirements are available as the test basis.

18.3.3 Test design

Test design is the activity that defines "how" the test conditions are to be tested. The result of this activity is a suite of test cases using the test techniques that have been selected in the test plan. Test cases may be designed one by one with the objective, when designing a new test case to be added to the suite, of covering at least one as yet uncovered test condition. Often, a new test case covers a few test conditions and, sometimes, the test analyst[1] has the possibility of choosing whether to address a single new test condition or several different ones with the test case under design. It must be noted that the latter possibility will improve the efficiency in any case in which the test is passed successfully. If, on the contrary, the system fails the test, it will be more complicated to establish the reason, there being several new test conditions as potential causes of the failure.

[1]The term *test analyst* is broadly used in the testing context and applies also to activities such as design and implementation.

In general, for test conditions specified at a higher level, it is more likely that test design is a separate activity. For lower levels of testing, test design may be conducted as an activity integrated with test analysis.

18.3.4 Test implementation

Test analysts implement, organize, and prioritize test cases during this activity. Test designs are implemented as "concrete test cases," test procedures, and test data. A test case is a set of input data, initial conditions, and expected results.

The implementation of a test case may require the test analyst to create some stored test data. For example, in order to test the condition T1 of the previous section, let us consider the test case <username: James.Marshall, password: JHExperience1966; expected result: "Access accorded">. This test case, however, will fail in the absence of certain stored data. Indeed, there must have been created a proper account for the user James Marshall prior to the execution of the test case. This requires that the data <James.Marshall, JHExperience1966> is stored in a file or a database.

Test cases may be dependent one on another. In such a case, a certain order of execution must be respected. A test case TCx may depend on TCy because this latter places the system in a final state, which is the one required as a precondition of TCx. Considering once again the test condition T1, if there is a function for the registration of the accounts to be tested, the test analyst can test this function with a test case like <username: James.Marshall, password: JHExperience1966; expected result: "Registration OK">. Next, she/he can test the condition T1 as performed previously, but without the need to store manually any data.

Dependent test cases are organized in test procedures that define lists of prioritized tests.

18.3.5 Test execution

Test execution can start when the entry criteria to test execution are satisfied. The entry criteria are the necessary conditions (e.g., the testable code is available, the requirements are specified and approved, the test cases are developed, etc.) that must hold before the test can be executed.

After having executed the test, it is necessary to compare the actual results with the expected results. A few kinds of problems may take place at this stage. In fact, a failure may be missed resulting, in a "false-negative" or a correct behavior may be misinterpreted with the consequence of registering a "false-positive."

In any case in which the actual and expected results are different, an "incident" is registered. Test analysts pay a lot of attention to the analysis of incidents to understand the cause, which might be a defect of the system, or a defect in the test objective. Testers prepare an incident report to provide any information useful for the isolation of the defect.

Low-level testing activities, such as unit testing and integration testing, are usually performed by testers and/or developers, whereas higher levels of testing (system testing and acceptance testing) may involve other stakeholders and the client. This is a way of building the confidence in the system.

18.3.6 Test evaluation exit criteria

From the point of view of the test process, test progress monitoring involves ensuring the collection of proper information to support the reporting requirements.

After having executed a test or group of tests, it is necessary to measure the progress toward completion. Typically, the completion of the testing activities is determined with respect to exit criteria which may be simple or more articulated conditions that combine constraints of the type "must" and "should." An example of an exit criterion established in the test plan is "There must be no high-risk issue open and there should be a 90% pass rate over all test cases." This means that the testing activities must continue until all high-priority risks have been mitigated with a global test pass rate greater than 90%. On the contrary, the testing activities could be terminated with a lower pass rate if all the high risks have been mitigated because, for example, the deadline for the testing activity has passed or the planned budget has been spent.

18.3.7 Test closure

This activity is performed once the testing execution exit criteria have determined that the objective has been achieved. At this stage, any relevant outputs and testing summaries are collected, analyzed, and archived. In particular, the tasks performed are (1) a test completion check, to ensure that all the test work has actually been concluded. As an example, for each planned test it must be verified if it has either been executed or deliberately skipped; (2) a test artifact handover, to distribute valuable work products to other participants who need them. For example, a list of known unsolved defects, along with a record of regression testing, should be provided to staff responsible for the maintenance of the software. A list of unsolved defects should also be provided to system users; (3) a lessons learned list, to perform retrospective meetings where important lessons can be documented to allow for good practices to be repeated and poor practices to be avoided in the future; and (4) an archiving of the results, to store any potentially useful information. This could include the test suite and the environment linked to the system and its version, the test plan, and the residual risks, etc.

18.4 Test metrics

Metrics represent one of the best tools to monitor the execution of a process and the degree of success achieved.

Testing metrics are categorized as

- Project metrics, which measure the progress of the testing activities with respect to the planned exit criteria;
- Process metrics, which measure the effectiveness of the testing activities, such as the percentage of defects detected by testing;
- Product metrics, which measure the characteristics of the product, such as, for example, the defect density; and
- People metrics, which measure the capability of individuals or teams.

Table 18.1 Test progress metrics

Dimension	Metric
Product quality risks	Percentage of risks completely covered by passing tests Percentage of risks for which some or all tests fail Percentage of risk not yet completely tested Percentage of risks covered, sorted by risk category Percentage of risks identified after the initial quality risk analysis
Defects	Cumulative number reported (found) versus cumulative number resolved (fixed) Mean time between failure or failure arrival rate Breakdown of the number or percentage of defects categorized by particular test items or components, root causes, source of defect (e.g., requirement specification, new feature, regression, etc.), test releases, priority/severity, reports rejected or duplicated, and phase introduced, detected, and removed Trends in the lag time from defect reporting to resolution Number of defect fixes that introduced new defects (sometimes called daughter bugs)
Tests	Total number of tests planned, specified (implemented), run, passed, failed, blocked, and skipped Regression and confirmation test status, including trends and totals for regression test and confirmation test failures Hours of testing planned per day versus actual hours achieved Availability of the test environment (percentage of planned test hours when the test environment is usable by the test team)
Convergence	Requirements and design elements coverage Risk coverage Environment/configuration coverage Code coverage

Many metrics may fall into more than one of these categories.

The use of testing metrics enables the Test Manager to collect results and report them to other stakeholders in a consistent way and enables a coherent tracking of progress over time. Metrics may be adopted to determine the overall success of a project. Therefore, great attention should be taken when selecting what to track and how often to report it. In particular, a Test Manager in the definition of the metrics should select a limited set of really useful metrics depending on the specific objectives of the project, process, and/or product. The set of chosen metrics should provide a balanced view since a single metric may give a misleading impression of the status or trends. The selected metrics should be defined in a clear and unambiguous way in order to avoid misinterpretations among the stakeholders. After a set of metrics has been defined, the tracking activity may start. Reporting should be as automated as possible and any divergence in the actual measurements from the expectations should be carefully analyzed and documented along with the reasons. Any reports should provide an immediate understanding of the status and trend, for management purposes.

Table 18.2 Process monitoring metrics

Metrics to monitor test planning and control
Risk, requirements, and other test basis element coverage
Defect discovery
Planned versus actual hours to develop testware and execute test cases

(a) Test planning and control

Metrics to monitor test analysis
Number of test conditions identified
Number of defects found during test analysis (e.g., by identifying risks or other test conditions using the test basis)
Planned versus actual hours to develop testware and execute test cases

(b) Test analysis

Metrics to monitor test design
Percentage of test conditions covered by test cases
Number of defects found during test design (e.g., by developing tests against the test basis)
Planned versus actual hours to develop testware and execute test cases

(c) Test design

Metrics to monitor test implementation
Percentage of test environments configured
Percentage of test data records loaded
Percentage of test cases automated

(d) Test implementation

Metrics to monitor test execution
Percentage of planned test cases executed, passed, and failed
Percentage of test conditions covered by executed (and/or passed) test cases
Planned versus actual defects reported/resolved
Planned versus actual coverage achieved

(e) Test execution

Metrics to monitor test progress and completion
Number of test conditions, test cases, or test specifications planned and those executed broken down by whether they passed or failed
Total defects, often broken down by severity, priority, current state, affected subsystem, or other classification
Number of changes required, accepted, built, and tested
Planned versus actual cost
Planned versus actual duration
Planned versus actual dates for testing milestones
Planned versus actual dates for test-related project milestones (e.g., code freeze)
Product (quality) risk status, often broken down by those mitigated versus unmitigated, major risk areas, new risks discovered after test analysis, etc.
Percentage loss of test effort, cost, or time due to blocking events or planned changes
Confirmation and regression test status

(f) Test progress and completion

(*Continues*)

Table 18.2 (Continued)

Metrics to monitor test closure
Percentage of test cases executed, passed, failed, blocked, and skipped during test execution
Percentage of test cases checked into re-usable test case repository
Percentage of test cases automated, or planned versus actual test cases automated
Percentage of test cases integrated into the regression tests
Percentage of defect reports resolved/not resolved
Percentage of test work products identified and archived
(g) Test closure

Table 18.3 Defect attributes

Attribute	Definition
Defect ID	Unique identifier for the defect
Description	Description of what is missing, wrong, or unnecessary
Status	Current state within defect report life cycle
Asset	The software asset (product, component, module, etc.) containing the defect
Artifact	The specific software work product containing the defect
Version detected	Identification of the software version in which the defect was detected
Version corrected	Identification of the software version in which the defect was corrected
Priority	Ranking for processing assigned by the organization responsible for the evaluation, resolution, and closure of the defect relative to other reported defects
Severity	The highest failure impact that the defect could (or did) cause, as determined by (from the perspective of) the organization responsible for software engineering
Probability	Probability of recurring failure caused by this defect
Effect	The class of requirement that is impacted by a failure caused by a defect
Type	A categorization based on the class of code within which the defect is found or the work product within which the defect is found
Mode	A categorization based on whether the defect is due to incorrect implementation or representation, the addition of something that is not needed, or an omission
Insertion activity	The activity during which the defect was injected/inserted (i.e., during which the artifact containing the defect originated)
Detection activity	The activity during which the defect was detected (i.e., inspection or testing)
Failure reference(s)	Identifier of the failure(s) caused by the defect
Change reference	Identifier of the corrective change request initiated to correct the defect
Disposition	Final disposition of defect report upon closure

According to the ISTQB Advanced Level Syllabus for Test Managers [53], there are five primary dimensions on which test progress may be monitored: Product (quality) risks, Defects, Tests, Coverage, and Confidence. Table 18.1 reports the metrics introduced for the first four dimensions, whereas the degree of confidence can be evaluated through surveys.

Table 18.4 Failure attributes

Attribute	Definition
Failure ID	Unique identifier for the failure
Status	Current state within failure report life cycle
Title	Brief description of the failure for summary reporting purposes
Description	Full description of the anomalous behavior and the conditions under which it occurred, including the sequence of events and/or user actions that preceded the failure
Environment	Identification of the operating environment in which the failure was observed
Configuration	Configuration details including relevant product and version identifiers
Severity	As determined by (from the perspective of) the organization responsible for software engineering
Analysis	Final results of causal analysis on conclusion of failure investigation
Disposition	Final disposition of the failure report
Observed by	Person who observed the failure (and from whom additional detail can be obtained)
Opened by	Person who opened (submitted) the failure report
Assigned to	Person or organization assigned to investigate the cause of the failure
Closed by	Person who closed the failure report
Date observed	Date/time the failure was observed
Date opened	Date/time the failure report is opened (submitted)
Date closed	Date/time the failure report is closed and the final disposition is assigned
Test reference	Identification of the specific test being conducted (if any) when the failure occurred
Incident reference	Identification of the associated incident if the failure report was precipitated by a service desk or help desk call/contact
Defect reference	Identification of the defect asserted to be the cause of the failure
Failure reference	Identification of a related failure report

Metrics, such as those reported in Table 18.2, can be defined to monitor the test process itself. In addition to such metrics, other project management techniques such as, for example, Work Breakdown Structures, are adopted to monitor the test process. In Agile frameworks, the testing progress is monitored through the burndown chart as any user story.

18.5 Defect management

Defect management begins after a tester has observed an anomaly and her/his preliminary analysis excludes the possibility of a false positive; i.e., the anomaly is a real system failure. It is worth noting that the process and actions described in the remaining part of this section do not apply in the case of Test Driven Development (TDD). Indeed, in TDD, the tests are written before the code so that the system is supposed to fail. In such a case, however, no incident is opened to find a defect, but the code is enriched to satisfy the test.

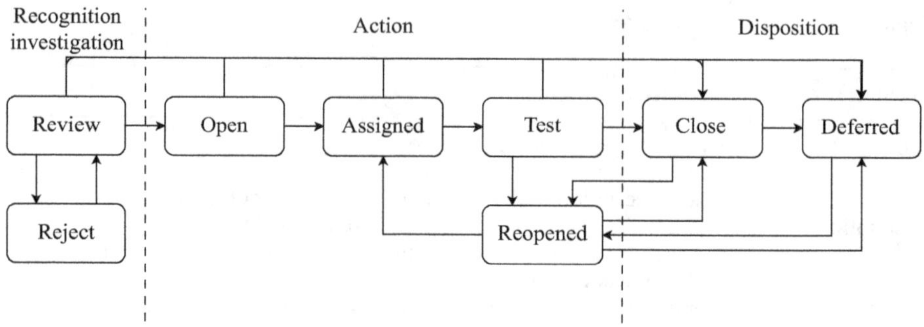

Figure 18.2 Defect management life cycle

It is possible to identify four steps in an incident management life cycle, namely:

1. Recognition, which occurs when the tester observes an anomaly, that is, a potential failure. Since testing activities, either static or dynamic, are extended throughout the entire development cycle, recognition can occur at any time;
2. Investigation, performed after the recognition, can reveal related issues and propose solutions. A possible conclusion is that the incident has not been caused by an actual defect;
3. Action, which is activated whenever the investigation has determined that the incident has been caused by an actual defect, which is necessary to remove. If the defect is resolved, regression testing and confirmation testing must be executed; and
4. Disposition, which consolidates the information gathered on the incident after its resolution and places the anomaly in a final status.

During the recognition step, the tester collects and records any preliminary supporting data. The anomaly is described through attributes, such as those reported in Table 18.3 that have been defined by IEEE 1044, the Standard Classification for Software Anomalies [54].

During the subsequent steps, such supporting data are integrated and refined. For example, in addition to the anomaly attributes, certain failure attributes (Table 18.4) can be collected.

Figure 18.2 reports an articulated defect life cycle management workflow, with several possible sequences of states for any defect, proposed in [55]. Indeed, for example, a defect may or may not be treated with an action depending on the decision to remove or to accept it. A treated defect could also be reopened for example because a confirmation test has failed.

Chapter 19

Change management, configuration management, and change management

19.1 Change management

According to IEEE 1044 Standard for Software Anomalies Classifications, there are three types of software change requests:

Corrective changes are those necessary to fix defects. This includes defects discovered by end users and those found through internal testing and other means. This kind of request is the one issued by the tester whenever the investigation of the anomaly concludes that it is a failure and there is a defect to correct.

Adaptive changes deal with adapting software to variations in the environment. The term environment refers to all possible external conditions that can influence the system. For example, the base hardware/software platform, any business rule or government policy, etc. The need for adaptive change emerges from monitoring the environment.

Perfective changes are those that improve a software component by adding new features or capabilities. This includes both changes due to new functional requirements and changes determined by new nonfunctional requirements (e.g., safety or security improvements).

In software maintenance, it is quite common to identify another type of software change, namely **Preventive changes**, because code can have flaws that are not real defects (they do not cause failures), but they make the software harder to maintain in the future. Preventive changes are those that improve the maintainability of a software system or reduce the potential for future failures.

The distribution of effort in an organization among the different maintenance operations is shown in Figure 19.1.

ISO/IEC 14764 (Standard for Software Engineering—Software Life Cycle Processes—Maintenance) [56] defines all four previous categories and classifies adaptive changes and perfective changes as enhancements, and corrective changes and preventive changes as corrections.

Such a standard defines a change (maintenance) process that consists of six phases:

1. Maintenance implementation. During this phase, whoever has the responsibility at maintaining the system (the Maintenance Manager) determines the plan and

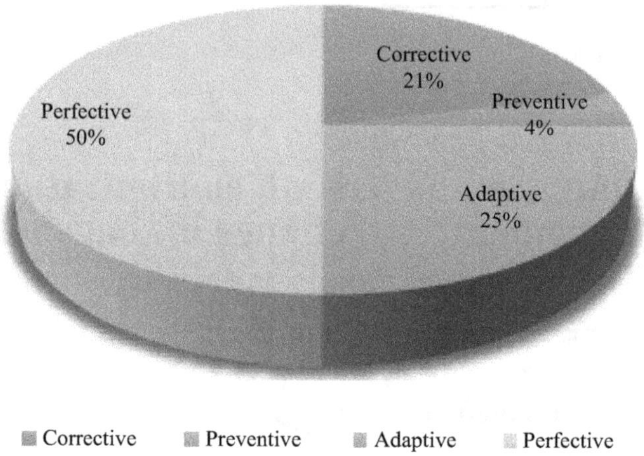

Figure 19.1 Distribution of maintenance effort

procedures to be executed during the Maintenance Process. The maintenance plan should be developed contemporaneously with the software development plan. A strategy for performing a software change should be defined and documented. The list of outputs of this activity includes: a maintenance plan, any maintenance procedures, a training plan, a maintenance manual, any problem resolution procedures, a measurement plan, and a configuration management plan.

2. Change request. This is the phase in which a change request is expected.
3. Impact analysis. After a change request has been received, the maintainer analyzes the change request and/or problem report, replicates and/or verifies the problem (if any), sets options for implementing the modification, documents the change request and/or problem report, the results and execution options, analyzes the impact of the requested change, and obtains approval for the selected modification option.
4. Change implementation. During the change implementation phase, the maintainer develops and tests the modification of the software product.
5. Change review/acceptance. This activity reviews the change in order to ensure that the modification to the system is proper and that it was performed in accordance with the approved standards using the correct methodology. The phase concludes with the acceptance/rejection of the modification.
6. Migration. A system may have to be modified in order to run in a new environment. The maintainer has to establish the actions necessary to perform the migration and then develop and document such actions.
7. Retirement. The retirement of a software product also requires some maintenance operations. The impact analysis supports the decision to retire the product. This analysis should establish which is the most cost-effective option between retaining outdated technology in service and replacing it with a new software product.

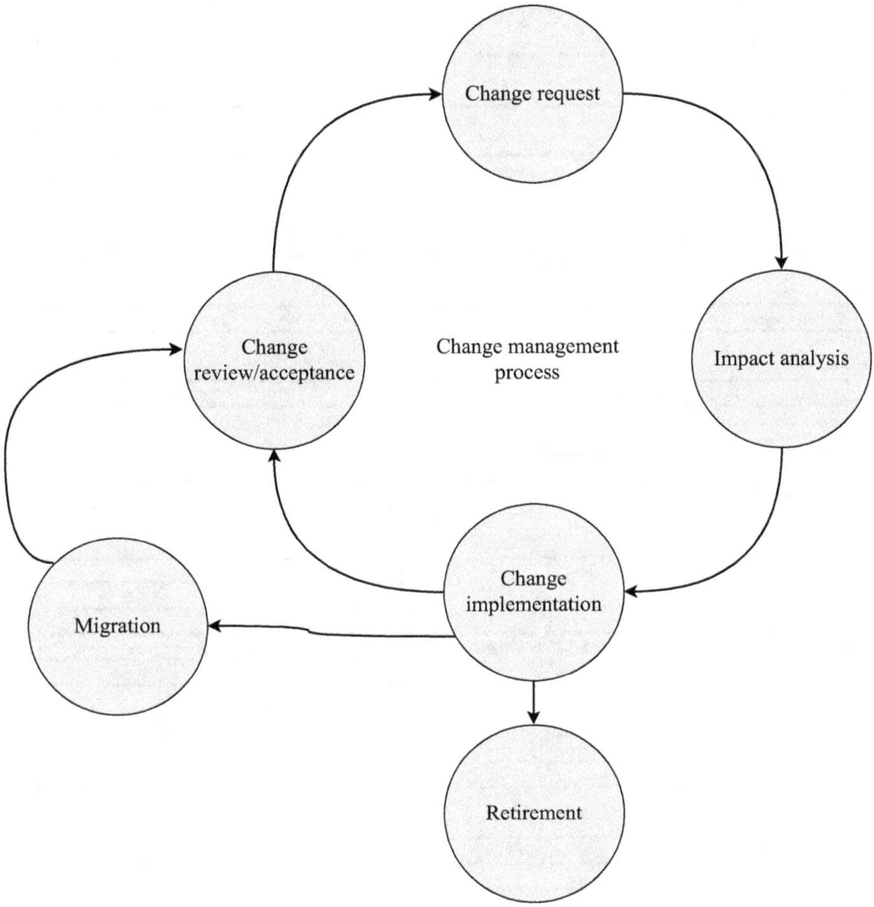

Figure 19.2 A change management process

Figure 19.2 shows an ISO/IEC 14764 compliant change management process. In this process, migrations and retirements are considered two special kinds of implementation. The former needs a successive review/acceptance like any other implementation, with the exception of a retirement that is exempted, the definitive decision about the retirement having already been being taken during the impact analysis.

It is finally useful to show a template for change requests (see Figure 19.3). This form comprises information that is collected at different times/stages of the change management process. Initially, only information about the type of request and the requestor included is in the document. After the impact analysis has been completed, the results are documented in the technical evaluation and time estimation sections. Finally, the decision is recorded in the approval/rejection section.

Title of Request:			
Request ID:			
Originator:			
Submission date:			
Change type			
Corrective		☐ Perfective	
Adaptive		☐ Preventive	
Summary Of the Problem			
Preliminary Cause Analysis			
Proposed Change Description			
Impact Analysis			
Directly changed module(s):		1.	
		2.	
		…	
Dependent Module(s):		1.	
		2.	
		…	
Estimated scope of source code changes:		**Lines of source code**	
		Change	
		Add	
		Delete	
Regression Test Plan			
Test suite:			
Test cases:		**New**	
		Modified	
Effort estimatation in hour per phase			
Requirements			
Analysis			
Design			
Implementation			
Testing			
….			
TOTAL ESTIMATED EFFORT			0,00
Approval/Rejection			
Approver			
Data			
Reason			
FINAL DECISION			

Figure 19.3 A software change request form

19.2 Configuration management

Configuration management is “(1) a discipline applying technical and administrative direction and surveillance to: identify and document the functional and physical characteristics of a configuration item (CI), control changes to those characteristics, record and report change processing and implementation status, and verify compliance with specified requirements (2) technical and organizational activities comprising configuration identification, control, status accounting, and auditing” [57].

Configuration management concerns the method of processing system changes systematically with the primary aim of realizing a system modification without affecting its integrity. To achieve this goal, detailed policies and procedures are necessary to manage revisions, track their status, and verify and document each step of the process.

A CI is an asset designated to be under the control of the configuration management process. It is treated as a single entity throughout the process. CIs may vary widely in complexity, size and type, ranging from an entire system including all hardware, software and documentation, to a single module or a minor hardware component.

A configuration is the set of functional and physical characteristics of a final deliverable.

Configuration management can also be regarded as an asset control tool, which is essential if multiple versions of a deliverable are to be created. The simplest kind of configuration management is version control.

By providing control of the project deliverables, configuration management avoids mistakes and misunderstandings.

This process is closely related to change management. These two processes, indeed, integrate with each other to ensure that any project deliverable meets the required specification, any change is beneficial, and there is a complete audit trail for the development of each deliverable.

The majority of the configuration management process is conducted at project level because that is where the deliverables are produced.

In Agile projects, the initial configuration is typically very flexible and updated frequently, such as, for example, in the case of the Continuous Integration approach.

An example of a configuration management process is shown in Figure 19.4. It consists of five activities:

1. Configuration management planning, which focuses on the realization of a configuration management plan. The plan should specify any specific procedures and the scope of their application during the life cycle. It should also identify roles and responsibilities for the carrying out of configuration management activities;
2. Configuration identification, which deals with breaking down the work into component deliverables (CIs) identified by a unique numbering system and defining configuration baselines;
3. Configuration control, which guarantees that all changes to CIs are documented and CI dependencies are managed properly;

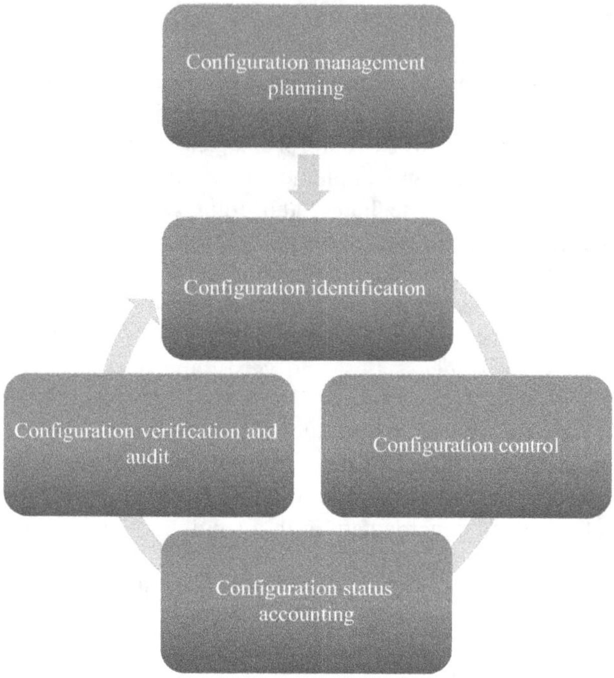

Figure 19.4 A configuration management process

4. Configuration status accounting, which monitors the current status of a configuration and provides traceability of those CIs that are under development; and
5. Configuration verification and audit, which establishes whether or not a deliverable conforms to its requirements and configuration policies and constraints. A configuration audit may take one of three forms: physical, functional, or system. A physical audit checks for the relevant elements of a CI and confirms whether or not the item meets its requirements. A functional audit of a CI verifies whether or not it performs the function for which it has been developed. A system audit checks whether or not the configuration management system is working and provides support for the quality system.

Figure 19.5 shows the table of contents of a configuration management plan.

The purpose of the configuration management plan is described in the first section of the document. This specifies how configuration management is managed throughout the project life cycle and how CI changes are performed and communicated to all interested stakeholders.

Roles and responsibilities are established in the configuration management plan. In the example, the Configuration Control Board, which may be composed of the Project Manager, Project Sponsor, Configuration Manager, and the team responsible for the configuration management, is supposed to review and approve/reject any

Figure 19.5 A configuration management plan table of contents

configuration change requests and assure that all approved changes are stored in the configuration management database. The Project Manager has the overall responsibility for all configuration management activities and must identify the CIs. The Configuration Manager has the overall responsibility for managing the configuration management database, establishing configuration standards and document templates, and providing any required configuration training.

The configuration management database stores the configuration information. This contains not only the configuration information for any asset (CI), but also related information such as the physical location of the item, its ownership, and its relationship to the other CIs.

Accounting for the status of the configuration requires the collection, processing, and reporting of the configuration data for all the identified CIs.

The configuration management plan should also establish the policies and strategies adopted for the configuration audits.

Finally, the Configuration Control section describes the process of systematically controlling and managing all steps of the configuration throughout the project life cycle.

19.3 Incident management

The regulations require that manufacturers conduct postmarket surveillance. In particular, the EU Directive 93/42/EEC establishes that any manufacturer must institute and maintain an up-to-date procedure to review the experience gained from the product use in the postproduction phase and must implement any appropriate means to apply necessary corrective actions.

Postmarket surveillance is a process adopted to monitor the performance of a medical device, and whose activities are designed to produce information regarding the use of the device to identify potential or occurring issues. The postmarket surveillance process must be activated immediately upon commercialization of the product.

Actions undertaken by the manufacturer during the postmarket surveillance may be proactive or reactive. In the case of a software product, a possible proactive action is the gathering of customer surveys. For example, a usability questionnaire can reveal potential usability issues and human errors.

In order to be reactive to incidents occurring in use, instead, the manufacturer has to implement an Incident Management process, which is strictly related to other processes such as change management—the resolution of an incident may require the raising of a change request—and configuration management.

Incident Management requires a "service desk" also known as the "help desk," which is the point of contact for all users communicating an incident or any condition that has the potential to result in a degradation of the quality of service.

Service desk personnel implement the level one support, which includes activities such as incident identification, logging, categorization, prioritization, and diagnosis. Other activities that may (not necessarily) be transferred to level 2 support are incident resolution and closure, and the communication of the result to the users.

Although in general Incident Management is not expected to perform a root cause analysis to identify why an incident has occurred and the focus is on doing whatever is necessary to restore the service, this is no longer applicable to Software as a Medical Device where an accurate investigation may be necessary before an attempt is made to fix the problem.

Figure 19.6 shows the ITIL Incident Management process [58], which consists of the following main steps:

- Incident identification. Incidents are communicated by users through any available channel. In the proposed workflow, there are four possible sources of incident communications that the manufacturer should support. Next, the service desk performs a quick analysis to establish whether or not the issue is truly an incident. Any other kind of request is categorized and handled differently from an incident.
- Incident logging. If identified as an incident, the service desk logs the event and opens a ticket. The ticket should include information, such as that necessary for a change request.
- Incident categorization. Categorization concerns the assignment of a category and one or more subcategories to the incident. In this way, some incidents are automatically prioritized and tracked accurately. When incidents are categorized, it is simple to collect statistics about, for example, the frequency of occurrence of certain incidents.
- Incident prioritization. The incident's priority is established by considering its impact on users and its urgency; that is, how quickly a resolution is necessary.
- Incident response. Incident resolution includes certain actions, such as Initial diagnosis, which takes place after the user has described her/his negative experience; Incident escalation, which occurs when an incident requires a second-level intervention; Investigation and diagnosis, which is needed to enable the staff to

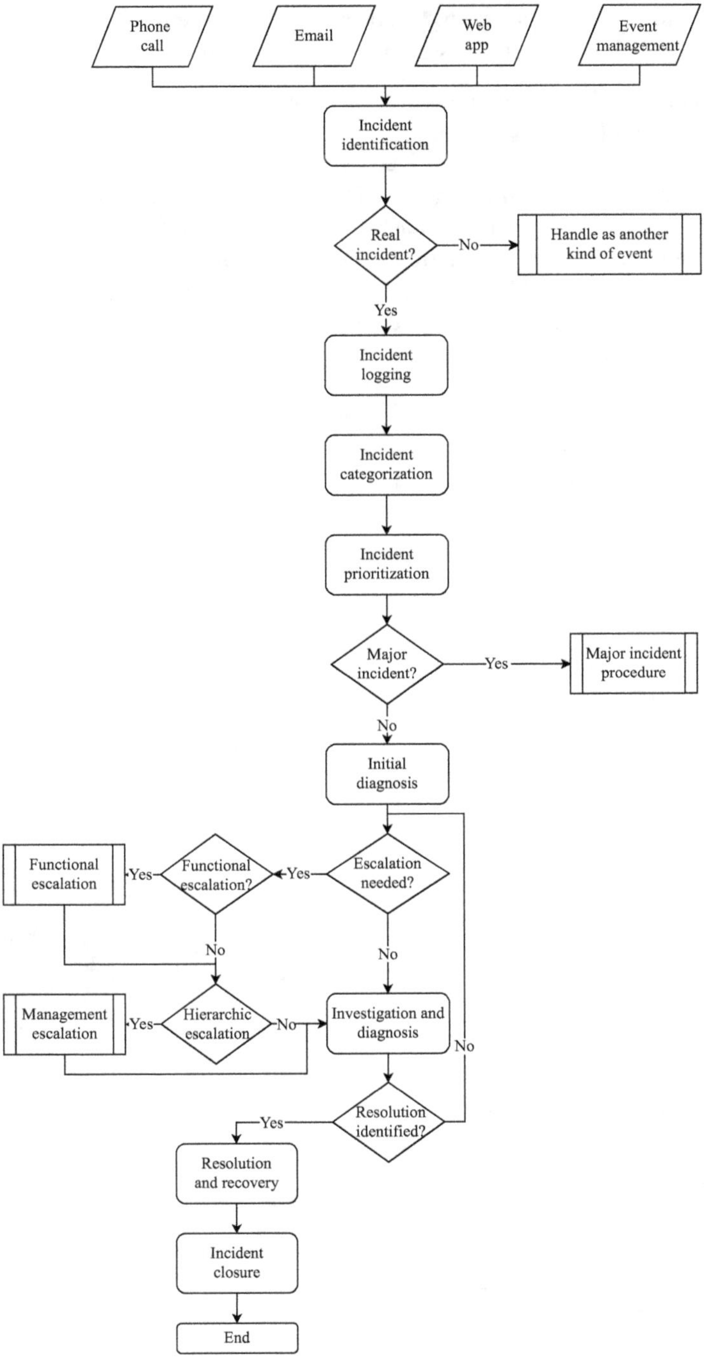

Figure 19.6 The ITIL Incident Management workflow

apply a solution; Resolution and recovery, which happens when the criticality has been solved; and Incident closure, which archives the issue.

An incident is labeled with one of the possible states at any given time of the Incident Management process:

"New," the service desk has received the incident, but it has not yet been assigned to an operator; "Assigned," the incident has been assigned to a service desk operator; "In progress," the incident has been assigned, but it has not yet been resolved; "On hold" or "pending," more information is needed to solve the issue; "Resolved," the incident is resolved and the user's functionality has been restored to the expected level of quality; and "Closed," the incident has been resolved and no further action is required.

Part VI

Conclusions

Chapter 20
Conclusions

20.1 Perspectives

The global market for Information Technologies in health care is experiencing and will continue to experience a dramatic development according to the predictions of the most quoted study centers. *bbcResearch* [59] evaluates the Compound Annual Growth Rates (CAGR) for IT spending to increase until 2019 by 4.8%, whereas *MedGadget* estimates a CAGR of 6.01% [60] (Figure 20.1).

Surfing among the pages of *bbcResearch*, it is possible to collect a large amount of data related to the segmented market, such as

"The global market for selected health self-monitoring technologies reached nearly $16.7 billion in 2016. This market should grow at a CAGR of 28.3% from $20.7 billion in 2017 to reach $71.9 billion by 2022:

- the global surgical robotics and computer-assisted surgery market reached nearly $3.5 billion in 2015. This market is expected to increase from $4.0 billion in 2016 to $6.8 billion in 2021 at a CAGR of 11.3% for 2016–21;
- the global mHealth market should reach $46.2 billion by 2021 from $13.2 billion in 2016 at a CAGR of 28.6%, from 2016 to 2021;
- the global market for wearable medical devices was valued at $4.8 billion in 2015. This market is expected to increase from $5.5 billion in 2016 to nearly $19.5 billion in 2021 at a CAGR of 28.8% for 2016–21;
- the global market for medical device technologies reached $458.3 billion in 2015. The market should reach $483.5 billion in 2016 and $634.5 billion by 2021, growing at a CAGR of 5.6% from 2016 to 2021;
- the global market for elderly care technology products was valued at $3.7 billion in 2014. This market is expected to grow to $10.3 billion in 2020 from nearly $4.4 billion in 2015, at a CAGR of 18.8% from 2015 to 2020;
- the US market for assistive technologies is projected to grow from $40.6 billion (including eyeglasses and contact lenses) in 2014 to $43.1 billion in 2015 and $58.3 billion in 2020, with a CAGR of 6.2% between 2015 and 2020;
- the global market for health-care analytics reached $4.8 billion in 2014. This market is forecasted to grow at a CAGR of 24.0% to reach nearly $16.9 billion in 2020 from nearly $5.8 billion in 2015; and

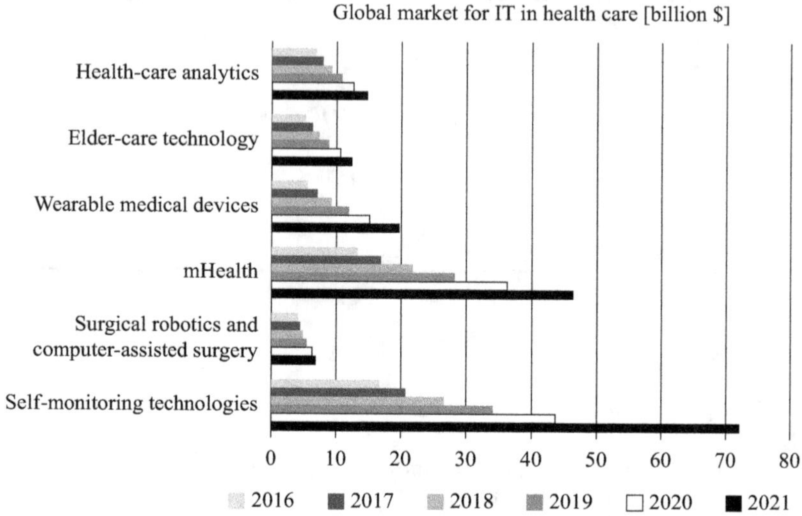

Figure 20.1 The global market for IT in health care

- the US elderly care market reached $381.6 billion in 2014 and nearly $417 billion in 2015. The market should reach $512.7 billion in 2020, demonstrating a CAGR of 4.2% from 2015 to 2020."

Another remarkable index of the growth of the business is the global Venture Capital funding for health-care IT/Digital Health companies that, according to the *Mercom Capital Group*, in the first quarter of 2017 almost doubled with $1.6 billion compared to $845 million in the fourth quarter of 2016 [61]. The funded areas in the first quarter of 2017 were Appointments reservation $315M, Mobile Wireless $230M, Data Analytics $193M, Population Health Management $115M, Telemedicine $112M, and the Social Health Network $102M (Figure 20.2).

On the other hand, technology innovation will show further signs of disruption in the next few years.

At the beginning of 2017, Forbes stated that 2017 will continue to be a year of tumultuous uncertainty and turbulence, but technology will continue to flourish and will have an unprecedented impact toward a connected home and health-care ecosystem [62].

The following are some of their top predictions for health care in 2017:

"**Blockchain becomes one of the most important technologies in the health-care industry**. With the potential to change how health-care information is stored, shared, secured, and paid for, blockchain technologies have immense potential to tackle some of the biggest challenges in health-care information management.

Artificial intelligence (AI) transforms medical imaging informatics. As more and more experts and health-care professionals find these AI systems usable as decision

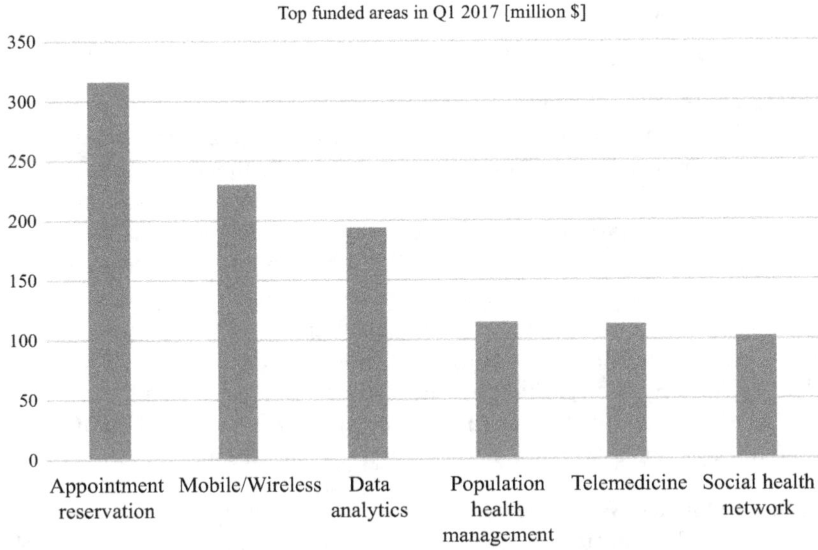

Figure 20.2 The top funded areas

support tools and not decision makers, uptake of AI-enabled clinical decision support tools is expected to increase in the coming years. More particularly during 2017, AI will play a big role in diagnostic imaging by complementing radiology with advanced interpretation and imaging informatics supports.

The digital health toolkit comes to behavioral health. Digital health coaching platforms and wellness programs with proven behavioral therapies will find their place as an efficient alternative to postcare settings and rehabilitation centers. Innovative online patient engagement platforms are capable of capturing tailored information on lifestyle and behavioral health. This is based on health risks data that will have a white space opportunity to provide patient risk classification solutions to make precision medicine a holistic approach.

With the aim of avoiding excessive future treatment costs, organizations will encourage healthier lifestyles among their members; they are likely to provide them with wearables and incentives to attain specific health goals as motivation. In the new year, well-being programs will become a central, critical business imperative, necessary to optimize the productivity and performance of employees, but also to manage the bottom line.

Point-of-care diagnostic devices push telehealth beyond video conferencing. Consumers will play a greater role in driving the uptake of point of care testing. In vitro diagnostic device manufacturers will invest in their digital strategy. This is in order to make their business models patient centric with consumer-friendly devices, to embed remote connectivity features for real-time access to data, and to simplify the sample collection process."

Frost & Sullivan's Visionary Health-care research has identified several technologies that are most likely to impact on health-care paradigms by 2025 [63]:

"**Quantum computing**. We are now beginning to see larger datasets in health-care research and delivery to analyze and make sense of entire genome sequences, the impact of environmental, behavioral, and hereditary factors on health, population health data, patient generated health data, etc. The amount of such data becoming available is only set to increase exponentially by 2025. The available computing prowess, even those of supercomputers, will not be adequate to generate quick and actionable insights from such large data sets. But quantum computing, that has a far greater calculation capacity than traditional computers, could help solve some of the highly complex health-care problems. One noteworthy company in this field is Canadian D-Wave Systems, which boasts of clients like NASA and Google. However, the possibility of widespread quantum computing is prevented by the problem of quantum incoherence, which, it is hoped, will be solved sometime soon.

Artificial intelligence. While the human capacity to analyze and make deductions is superior to any other species on the planet, it is still limited in terms of the volume of information that can be processed quickly. AI makes this process faster by several degrees and far more efficient than humanly possible. IBM's Watson, for example, can read 40 million documents in 15 s. With machine learning capabilities, the technology's health-care applications are boundless. Some of the applications currently being developed are assisting physicians and radiologists to make accurate diagnoses (IBM Watson Health), predicting which potential therapeutic candidates are most likely to work as efficient drugs (Atomwise) and mining medical records data to improve health-care service delivery (Google DeepMind Health).

Robotic care. Robots have been in health care for a long time now—the Da Vinci surgical robot is a case in point. But several other robotic applications are emerging and we should expect a lot more robots operating in the health-care space by 2025. Consider the simplistic telepresence robots like those offered by InTouch Health, allowing the doctor to 'move around' and examine patients, while being seated at his or her computer at a distant location. Or Aethon's TUG robots that help hospitals internally transport their pharmacy supplies, lab samples, patient food, clean or soiled linen or even trash, all by itself. Then, there are the patient and elderly care robots that help in lifting patients from beds to wheelchairs and back, like the Robear robot or the Riba robot in Japan. Finally, robots can also play a role in pediatric therapy for autism disorders, phobias and as distractions; several examples exist—Phobot, PARO, NAO, and Milo.

Nanorobots. At the nanoscale, robots can play entirely different roles, this time inside the human body, traveling through bloodstreams. Ongoing research is exploring the potential nanorobots can have in vitals monitoring, performing body functions (e.g., carrying oxygen, destroying infectious agents like bacteria), targeted drug delivery (e.g., cancer therapy, blood clotting), or even to perform nanoscale, in situ surgeries. The actual list of applications of nanomedicine, the umbrella term for nanotechnology applications in health care, is even larger and fascinating. It includes assisting biological research (cell simulations), being intracellular sensors for diagnostics and playing a role in molecular medicine (genetic therapy). At the very least, we should see the beginning of testing of such applications by 2025.

Cyborgization. The year 2025 should bring not just the introduction of robots inside our bodies, but also the transformation of the human body itself into partial robotic beings. This can manifest in several forms, some of which are visible even today—limb replacements, organ replacements, internal electronics, and capabilities, limbs, or senses that are enhanced in function than their normal counterparts. Apart from the 'bionic' prosthetics movement, an estimated 30,000–50,000 people already have an implanted Radio-frequency Identification (RFID) chip inside their bodies. In the future, we are likely to see enhanced capabilities in terms of vision, hearing or with limbs, especially in defense application areas. Artificial pancreas is a subject of intense research, and it is likely that more sophisticated versions of these devices may even be implanted in the human body in the future—to supplement or even completely replace normal pancreas.

Brain–computer interfaces. Another form of cyborgization is the use of brain–computer interfaces to connect a wired brain directly with an external device. Apart from the research and brain-mapping applications currently in use, the technology is being developed for 'neural bypass' applications—helping paralyzed patients regain control of their limbs via 'external' connections to the limbs. Similar applications are being developed wherein the body's neural framework is tapped using electric stimulation to modify certain functions. Existing examples include cochlear implants and pacemakers, while applications being developed include retinal implants (to restore sight) and spinal cord stimulators (for pain relief).

Medical Tricorder (diagnostic device). Taking cue from the device popularized by the Star Wars franchise, efforts are aimed at developing a hand-held portable diagnostic device that can scan the human body and diagnose their ailments within seconds. While the fantasy version of the device could do this, current efforts are more realistic in their approach. The $10 million Qualcomm Tricorder X Prize competition launched in 2012, for example, aims at diagnosing 13 medical conditions (ten required, three elective) including strep throat, sleep apnea and atrial fibrillation, with a consumer-friendly interface device weighing no more than 5 pounds. With the winners of this competition set to be announced in 2017, we could expect such devices to be commercially available by 2025.

Digital avatars. After self-diagnosing using a tricorder, patients in 2025 will want to get in touch with a doctor. Of course, telehealth will be an option, but there might be another option available for satisfying queries or getting more information on the diagnosis—just like the generic voice assistants available today. While Siri or Cortana are voice-only assistants, the Dr. WebMD of 2025 can be a digital avatar that can appear in holographic projections to assist patients and caregivers with their health-care queries. The holographic projection of a human doctor, backed by AI technologies, will allow for it to handle several queries simultaneously. Beyond answering queries, it could schedule appointments for a physical checkup with a doctor in your network, and share notes of your conversation with a doctor, in a digital-physical care coordination model.

Augmented/virtual reality. The applications of the two related technologies are manifold and relevant to both sides of the care delivery equation—providers as well as patients. Providers can benefit from using enabled glasses for medical education—to study the human anatomy, for example, and for observing and studying surgeries as they were performed. Augmented reality could also be used during live surgeries to 'see through' anatomical structures to know the location of organs and blood vessels. On the patient side, one of the most advanced applications that are already in use is the treatment of various phobias and other mental health disorders. As the technology advances, we can expect more advanced applications to emerge by 2025, especially for health-care providers.

3D printing. 3D printing is a well-known technology with several existing applications in health care, including orthopedic devices and several implants. Another application that is being considered is of 3D printed medicines, which can allow alteration of daily dosage and enable personalized medicine by customizing formulations of the drugs. Another niche that is now opening up is that of 3D Bioprinting—the possibility of 'printing' tissues or even organs. Applications range from skin tissue for burn victims to organ replacements for patients. Tissues thus printed can also be used in drug development."

20.2 Criticality

Upcoming Information Technologies, as predicted by most analysts and shown in the previous section, will potentially have a disruptive impact on the way health-care services are realized by health-care organizations and provided to patients. The beneficial effects, however, may be dramatically reduced or impaired by the slow response of competent authorities in updating regulations with respect to the diffusion of such brand new technologies.

If we consider what has happened in Europe, the European Commission issued Directive 93/42 EEC in 1993. This Directive regulated tools adopted for medical practices and defined the term Medical Device formally. The early tools had already been realized and used by ancient civilizations. Archeological relics of ancient Rome, indeed, had been classified as tools for surgical operations. Therefore, the Directive finally regulated a market that had offered for a long time a variety of instruments adopted in medicine, for which no quality requirements had been formally demanded previously (Figure 20.3).

It is not surprising, however, that the term software is not found in the Directive. Indeed, such a regulation was conceived taking into account the traditional Medical Devices; i.e., hardware tools.

Only with the 2007 amendment to the Directive, did the term software finally appear in the definition of Medical Device. However, it mainly referred to embedded software and included only a few classes of stand-alone software systems such as, for example, medical imaging applications.

The delay in the consideration of Information Technology innovations is evidenced in this prediction of Mark Weiser. In 1991, 16 years before the term software

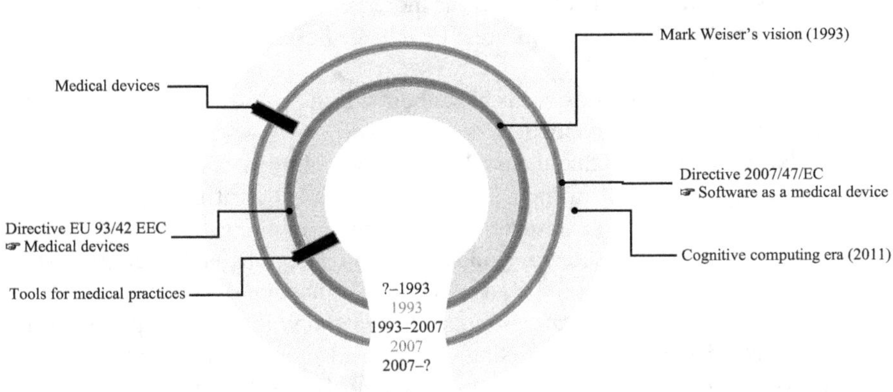

Figure 20.3 Technology innovation versus regulations timetable

appeared in the European regulations, Mark Weiser envisioned "The Computer for the 21st Century" [64] imagining a shift from the model One-Person-One-Computer to the model One-Person-Many-Computers, transparently distributed into our environments. Already in 1993, he coined the term Ubiquitous Computing. "Ubiquitous Computing has as its goal the enhancing of computer use by making many computers available throughout the physical environment, but making them effectively invisible to the user." It is the paradigm that has made possible the realization of Ambient Assisted Living applications, which are context-aware intelligent environments that support their users, especially elderly people, in daily living and working environments to stay active longer, remain socially connected and live independently. The first Ambient Assisted Living applications were made around the year 2000, 7 years before the term software was introduced (with a limited acceptance) into the definition of Medical Device.

Nowadays, we are witnessing another technology revolution: the cognitive computing era. A cognitive system is an AI-based system that performs the cognitive work of knowing, understanding, planning, deciding, solving problems, assessing, perceiving, and acting. A Cognitive computing system can analyze terabytes of data in seconds and make decisions driven by such data, rather than relying on instinct.

IBM's Watson—the first commercial Cognitive System—can learn, reason, and understand natural language. On February 2011, it participated and won the television quiz Jeopardy against two of the quiz's greatest champions. Successively, Watson went to medical school and has been studying oncology with the aim of improving the quality and speed of patient care through personalization. Andrew McAfee, co-director of the MIT Initiative on the Digital Economy and co-author of "The Second Machine Age," speaking about Watson declared "I am convinced that if it is not already the world's best diagnostician, it will be soon."

Watson is already capable of storing far more medical information than any doctor, and unlike humans, its decisions are all evidence based and free of any cognitive bias or overconfidence.

McAfee back in 2011 wrote in his blog about why a diagnosis from "Dr. Watson" would be a significant improvement:

"It is based on all available medical knowledge. Human doctors can't possibly hold this much information in their heads, or keep up it as it changes over time. Dr. Watson knows it all and never overlooks or forgets anything.

It is accurate. If Dr. Watson is as good at medical questions as the current Watson is at game show questions, it will be an excellent diagnostician indeed.

It is consistent. Given the same inputs, Dr. Watson will always output the same diagnosis. Inconsistency is a surprisingly large and common flaw among human medical professionals, even experienced ones. And Dr. Watson is always available and never annoyed, sick, nervous, hungover, upset, in the middle of a divorce, sleep-deprived, and so on.

It has very low marginal cost. It will be very expensive to build and train Dr. Watson, but once it is up and running the cost of doing one more diagnosis with it is essentially zero, unless it orders tests.

It can be offered anywhere in the world. If a person has access to a computer or mobile phone, Dr. Watson is on call for them."

The question now is, does Dr. Watson need an FDA approval before being trusted with its diagnoses? In the end, it is a software program. If so, is it certifiable?

It is difficult to think that it has been produced by following an IEC 62304 compliant development process. It is very unlike anything that has been realized under the umbrella of an ISO 13485 or FDA 820.30 Quality Management System. More important, even if we wanted to produce a new Dr. Watson fully compliant with the regulation requirements, are we sure that such requirements would be proper for the nature of the system we have in mind? After all, such a kind of system, as any Decision Support System (even those much less sophisticated), relies on "knowledge" (knowledge-bases, patterns extracted from data, processes mined, etc.). How can we verify and validate the functions that use such knowledge? Is software testing good for verifying and validating the machine's knowledge? Probably not. Therefore, there exists a gap between these technologies and the regulations and software engineering practices that do not allow the production of evidence of quality and the building of the confidence in such systems.

In the absence of improvements in both regulations and software engineering practices, we see two possible routes: either Dr. Watson finishes "its studies" and obtains a degree at medical school or it (and other types of decision support systems) is not used as a Medical Device and the legal responsibilities are assumed by clinicians.

This second possibility opens up another criticality that could be very dangerous for those systems that allow the patient the self-assess her/his health status. A significant example is represented by the myriad of mobile Apps available for skin monitoring. A recent study [65] has evaluated some of the (noncertified) applications available on the market and has demonstrated a percentage of 30% of false negatives;

that is, for 30% of cases for which a melanoma should have been diagnosed and removed, the app did not alert the user.

These Apps can be on the market because they include simple disclaimers like "The information contained in this application is for general informational purposes only and is not intended as, nor should ever be considered as, a substitute for professional medical advice." For this reason, these Apps are not Medical Devices.

From a legal point of view, everything is in order. The producer has no responsibility, which is assumed completely by the user. Unfortunately, very often, the user does not really or completely understand the meaning of this disclaimer. She/he is likely not to be aware of Medical Device regulations and tends to believe that an App will, simply because it is on the market, provide reliable information.

20.3 Conclusions

In this book, we have tried to provide an overview of the regulations, standards and software engineering practices that should be taken into account in order to realize high-confidence certifiable Software as a Medical Device. We have introduced the regulations currently in force in Europe and the USA, and have added some other regulations for other markets. All such regulations share common elements such as the need for a classification system, a Quality Management System, a Risk Management framework, and certain verification and validation activities. However, they differ to a certain degree in imposing a few different requirements, making it impossible for a manufacturer to realize a Software as a Medical Device through a unique framework for all markets. The regulations, moreover, have been originally focused on traditional Medical Devices with the consequence that the application to software is not always possible or direct.

There exist a variety of Standards and technical guidelines that, although they are not mandatory, may be useful for the realization of products with a high degree of quality. Once again, the possibility of implementing a Standard as it is depends, country by country, on the regulations in force. In some cases, a certain Standard is recognized and may be applied as it is. In other cases, such a Standard has been harmonized to the regulations. In other cases again, the same standard is not recognized at all.

We have also shown some techniques for software verification and validation, and, finally, we have introduced some of the most common practices adopted for the management software projects in medical applications, including the latest ones, such as Agile methods.

References

[1] (2013) Software as a medical device (samd): Key definitions. [Online]. Available: http://www.imdrf.org/docs/imdrf/final/technical/imdrf-tech-131209-samd-key-definitions-140901.pdf.

[2] ISO/IEC, *ISO/IEC 9126. Software Engineering – Product Quality*. ISO/IEC, 2001.

[3] N. G. Leveson, *Safeware: System Safety and Computers*. New York, NY, USA: ACM, 1995.

[4] N. G. Leveson and C. S. Turner, "An investigation of the therac-25 accidents," *Computer*, vol. 26, no. 7, pp. 18–41, July 1993.

[5] N. Leveson, "Medical devices: The therac-25."

[6] (2001) General principles of software validation; final guidance for industry and FDA staff. [Online]. Available: https://www.fda.gov/Regulatory Information/Guidances/ucm085281. htm#_Toc517237955.

[7] (1990) Directive 90/385/eec. [Online]. Available: http://eur-lex.europa.eu/legal-content/EN/TXT/?uri=CELEX:01990L0385-20071011.

[8] (1993) Directive 93/42/eec. [Online]. Available: http://eur-lex.europa.eu/legal-content/EN/TXT/?uri=CELEX:01993L0042-20071011.

[9] (1998) Directive 98/79/ec. [Online]. Available: http://eur-lex.europa.eu/legal-content/EN/TXT/PDF/?uri=CELEX:01998L0079-20120111&qid=1413308118275&from=EN.

[10] (2007) Directive 2007/47/ec. [Online]. Available: http://eur-lex.europa.eu/legal-content/EN/TXT/?uri=CELEX:32007L0047.

[11] Medical devices: Guidance document – qualification and classification of stand alone software. [Online]. Available: http://ec.europa.eu/growth/sectors/medical-devices/guidance_it.

[12] (2000) Recommendation NB-MED/2.5.1/Rec5 – technical documentation. [Online]. Available: https://dannorth.net/introducing-bdd/.

[13] FDA mission. [Online]. Available: http://www.fda.gov/AboutFDA/WhatWeDo/default.htm.

[14] Medical device data systems, medical image storage devices, and medical image communications devices. [Online]. Available: http://www.fda.gov/downloads/MedicalDevices/DeviceRegulationandGuidance/GuidanceDocuments/UCM401996.pdf.

[15] General wellness: Policy for low risk devices. [Online]. Available: http://www.fda.gov/downloads/MedicalDevices/DeviceRegulationandGuidance/GuidanceDocuments/UCM429674.pdf.

[16] Guidance for the content of premarket submissions for software contained in medical devices. [Online]. Available: http://www.fda.gov/downloads/MedicalDevices/DeviceRegulationandGuidance/GuidanceDocuments/ucm089593.pdf.

[17] General principles of software validation. [Online]. Available: http://www.fda.gov/downloads/MedicalDevices/DeviceRegulationandGuidance/GuidanceDocuments/ucm085371.pdf.

[18] Off-the-shelf software use in medical devices. [Online]. Available: http://www.fda.gov/downloads/MedicalDevices/ ... /ucm073779.pdf.

[19] Mobile medical applications. [Online]. Available: http://www.fda.gov/downloads/MedicalDevices/ ... /UCM263366.pdf.

[20] (2016) Therapeutic goods (medical devices) regulations 2002. [Online]. Available: https://www.legislation.gov.au/Details/F2016C00150.

[21] (2001) Resolution RDC no. 185. [Online]. Available: http://portal2.saude.gov.br/saudelegis/leg_norma_espelho_consulta.cfm?id=3727179&highlight=&tipoBusca=post&slcOrigem=0&slcFonte=0&sqlcTipoNorma=186&hdTipoNorma=186&buscaForm=post&bkp=pesqnorma&fonte=0&origem=0&sit=0&assunto=&qtd=10&tipo_norma=186&numero=185&data=%20&dataFim=&ano=&pag=1.

[22] (1998) Medical devices regulations – SOR/98-282. [Online]. Available: http://laws-lois.justice.gc.ca/eng/regulations/sor-98-282/FullText.html.

[23] Regulations for the supervision and administration of medical devices. [Online]. Available: http://eng.sfda.gov.cn/WS03/CL0755/.

[24] Regulatory framework for federal service for surveillance in healthcare. [Online]. Available: http://www.roszdravnadzor.ru/en.

[25] (2016) ISO 15489-1:2016 – information and documentation – records management – part 1: Concepts and principles. [Online]. Available: https://www.iso.org/standard/62542.html.

[26] (2015) Statement regarding use of ISO 14971:2007 "medical devices – application of risk management to medical devices". [Online]. Available: http://www.imdrf.org/docs/imdrf/final/procedural/imdrf-proc-151002-risk-management-n34.pdf.

[27] (2009) IEC 31010: Risk management Ñ risk assessment techniques. [Online]. Available: https://www.iso.org/standard/51073.html.

[28] (2006) IEC 60812: Analysis techniques for system reliability – procedure for failure mode and effects analysis (FMEA). [Online]. Available: https://webstore.iec.ch/publication/3571.

[29] (2006) IEC 61025: Fault tree analysis (FTA). [Online]. Available: https://webstore.iec.ch/publication/4311.

[30] (2003) Application of hazard analysis and critical control point (HACCP) methodology to pharmaceuticals. [Online]. Available: http://www.who.int/medicines/areas/quality_safety/quality_assurance/ApplicationHACCPMethodologyPharmaceuticalsTRS908Annex7.pdf.

[31] (2016) IEC 61882: Hazard and operability studies (HAZOP studies) – application guide. [Online]. Available: https://webstore.iec.ch/publication/24321.

[32] (2015) Statement regarding use of IEC 62304:2006 "medical device software – software life cycle processes". [Online]. Available: http://www.imdrf.org/docs/imdrf/final/procedural/imdrf-proc-151002-medical-device-software-n35.docx.

[33] "IEEE standard for system and software verification and validation," *IEEE Std 1012-2012 (Revision of IEEE Std 1012-2004)*, pp. 1–223, May 2012.

[34] T. J. McCabe, "A complexity measure," *IEEE Transactions on Software Engineering*, vol. SE-2, no. 4, pp. 308–320, December 1976.

[35] R. Black and J. L. Mitchell, *Advanced Software Testing – Vol. 3: Guide to the ISTQB Advanced Certification As an Advanced Technical Test Analyst*, 1st ed. Santa Barbara, CA: Rocky Nook, 2011.

[36] VPTag project. [Online]. Available: https://sourceforge.net/projects/vptag/?source=typ_redirect.

[37] *BS 7925-2 Software Component Testing Standard.*

[38] E. M. Clarke and E. A. Emerson, "Design and synthesis of synchronization skeletons using branching-time temporal logic," in *Logic of Programs, Workshop*. London, UK: Springer-Verlag, 1982, pp. 52–71. [Online]. Available: http://dl.acm.org/citation.cfm?id=648063.747438.

[39] G. J. Holzmann, "The model checker spin," *IEEE Transactions on Software Engineering*, vol. 23, no. 5, pp. 279–295, May 1997. [Online]. Available: http://dx.doi.org/10.1109/32.588521.

[40] L. Cardelli and A. D. Gordon, "Mobile ambients," *Theoretical Computer Science*, vol. 240, no. 1, pp. 177–213, 2000. [Online]. Available: http://dx.doi.org/10.1016/S0304-3975(99)00231-5.

[41] C. Baier and J.-P. Katoen, *Principles of Model Checking*. London, England: The MIT Press, 2008.

[42] M. Leucker and C. Schallhart, "A brief account of runtime verification," *The Journal of Logic and Algebraic Programming*, vol. 78, no. 5, pp. 293–303, 2009. [Online]. Available: http://www.sciencedirect.com/science/article/pii/S1567832608000775.

[43] (2015) Ichaos report. [Online]. Available: https://www.infoq.com/articles/standish-chaos-2015.

[44] The agile manifesto. [Online]. Available: http://http://agilemanifesto.org/principles.html.

[45] (2001) The agile manifesto. [Online]. Available: http://docplayer.net/2823746-The-agile-manifesto-august-2001.html.

[46] K. Schwaber and M. Beedle, *Agile Software Development with Scrum*, 1st ed. Upper Saddle River, NJ, USA: Prentice Hall PTR, 2001.

[47] D. North. (2006) Introducing BDD. [Online]. Available: https://dannorth.net/introducing-bdd/.

[48] "AMII TIR45:2012 guidance on the use of agile practices in the development of medical device software," Advancing Safety in Medical Technology, 2012.

[49] (1990) I610.12-1990 – IEEE standard glossary of software engineering terminology. [Online]. Available: https://standards.ieee.org/findstds/standard/610.12-1990.html.

[50] (1999) Software requirements specification (SRS) template. [Online]. Available: www.cse.msu.edu/ cse870/IEEEXplore-SRS-template.pdf.

[51] (1993) IEEE recommended practice for software requirements specifications. [Online]. Available: http://ieeexplore.ieee.org/document/392555/.

[52] Design control guidance for medical device manufacturers. [Online]. Available: https://www.fda.gov/downloads/MedicalDevices/DeviceRegulationandGuidance/GuidanceDocuments/ucm070642.pdf.

[53] (2012) Advanced level syllabus – test manager. [Online]. Available: http://www.istqb.org/downloads/send/10-advanced-level-syllabus-2012/54-advanced-level-syllabus-2012-test-manager.html.

[54] "IEEE standard classification for software anomalies – redline," *IEEE Std 1044-2009 (Revision of IEEE Std 1044-1993) – Redline*, pp. 1–25, January 2010.

[55] R. Black, *Advanced Software Testing – Vol. 2: Guide to the ISTQB Advanced Certification As an Advanced Test Manager*, 2nd ed. Santa Barbara, CA: Rocky Nook, 2014.

[56] (2006) ISO/IEC 14764:2006 software engineering – software life cycle processes – maintenance. [Online]. Available: https://www.iso.org/standard/39064.html.

[57] "IEEE standard for configuration management in systems and software engineering – redline," *IEEE Std 828-2012 (Revision of IEEE Std 828-2005) – Redline*, pp. 1–126, March 2012.

[58] G. Blokdijk and I. Menken, *Incident Management Best Practice Handbook: Building, Running and Managing Effective Support and Incident Tracking – Ready to Use Supporting Documents Bringing ITIL Theory into Practice.* London, UK: Emereo Pty Ltd, 2008.

[59] (2017) BBCresearch reports. [Online]. Available: www.bbcresearch.com.

[60] (2017) Medgadget reports. [Online]. Available: www.MedGadget.com.

[61] (2017) Mercom capital group reports. [Online]. Available: http://mercomcapital.com/healthcare-it-digital-health-has-strong-first-quarter-with-$1.6-billion-in-q1-2017-vc-funding-reports-mercom-capital-group.

[62] (2017) Forbes reports. [Online]. Available: https://www.forbes.com/sites/reenitadas/2017/01/06/9-top-healthcare-predictions-for-2017/#5de3c68f5e68.

[63] (2017) Frost&sullivan reports. [Online]. Available: http://medtechevents.blogspot.it/2016/10/healthcare-2025-ten-top-technologies.html.

[64] M. Weiser, "Human-computer interaction," R. M. Baecker, J. Grudin, W. A. S. Buxton, and S. Greenberg, Eds. San Francisco, CA, USA: Morgan Kaufmann Publishers Inc., 1995, ch. The Computer for the 21st Century, pp. 933–940. [Online]. Available: http://dl.acm.org/citation.cfm?id=212925.213017.

[65] Joel A. Wolf, Jacqueline F. Moreau, Oleg Akilov, *et al.*, "Diagnostic inaccuracy of smartphone applications for melanoma detection," *JAMA Dermatology*, vol. 149, no. 4, pp. 422–426, 2013. [Online]. Available: +http://dx.doi.org/10.1001/jamadermatol.2013.2382.

Index

www.ingramcontent.com/pod-product-compliance
Lightning Source LLC
LaVergne TN
LVHW011748200926
848484LV00003B/654